Project Earth Science:
Physical Oceanography

Revised 2nd Edition

Project Earth Science: Physical Oceanography

Revised 2nd Edition

by Alfredo L. Aretxabaleta, Gregg R. Brooks, and Nancy W. West

NSTApress

National Science Teachers Association

Arlington, Virginia

National Science Teachers Association

Claire Reinburg, Director
Jennifer Horak, Managing Editor
Andrew Cooke, Senior Editor
Wendy Rubin, Associate Editor
Amy America, Book Acquisitions Coordinator

Printing and Production
Catherine Lorrain, Director

National Science Teachers Association
Francis Q. Eberle, PhD, Executive Director
David Beacom, Publisher

Art and Design
Will Thomas Jr., Director
Tracey Shipley, Cover Design
Front Cover Photo: NOAA Photo Library
Back Cover Photo: NOAA Photo Library
Banner Art © Gerry Boughan / Dreamstime.com

Revised 2nd Edition developed, designed, illustrated,
and produced by
Focus Strategic Communications Inc.
www.focussc.com

Library of Congress Cataloging-in-Publication Data

Aretxabaleta, Alfredo L., 1975-
 Project earth science: physical oceanography / by Alfredo L.
Aretxabaleta, Gregg R. Brooks, and Nancy W. West.
    p. cm.
 Includes index.
 ISBN 978-1-936959-02-0 (print) -- ISBN 978-1-936959-99-0 (e-book) 1.
Oceanography--Experiments. 2. Oceanography--Study and teaching (Middle
school) I. Brooks, Gregg R. II. West, Nancy W., 1957- III. Title.

 GC31.5.A75 2011
 551.46--dc23

2011017996

Featuring SciLinks—a new way of connecting text and the Internet. Up-to-the minute online content, classroom ideas, and other materials are just a click away. For more information go to www.scilinks.org/Faq.aspx.

Table of Contents

Readings

Acknowledgments

A number of people contributed to the development of this new edition of *Project Earth Science: Physical Oceanography*. The volume began as a collection of Activities and Readings from Project Earth Science (PES), a teacher enhancement project funded by the National Science Foundation. Principal investigators for this project were Iris R. Weiss, president of Horizon Research, Inc.; Diana Montgomery, research associate at Horizon Research, Inc.; Paul B. Hounshell, professor of education, University of North Carolina at Chapel Hill; and Paul Fullagar, professor of geosciences, University of North Carolina at Chapel Hill.

Project Earth Science provided inservice education for and by middle school Earth science teachers in North Carolina. Activities and Reading in this book underwent several revisions as a result of suggestions provided by PES teacher-leaders, principal investigators, and the project staff and consultants. PES leaders made this book possible through their creativity and commitment to the needs of students and classroom teachers. Brent A. Ford and P. Sean Smith were the authors of the first edition and were part of the PES team.

The authors would like to thank the many people who contributed to this revised second edition. Alfredo L. Aretxabaleta thanks Chris Sherwood of the United States Geological Survey (USGS) and Raymond Schmitt of the Woods Hole Oceanographic Institution (WHOI) for their helpful suggestions. Gregg R. Brooks thanks Rebekka Larson of Eckerd College and University of South Florida for providing valuable information. Nancy W. West expresses her appreciation to Geoff Feiss of Quarter Dome Consulting, LLC for his ideas and advice.

We also thank Adrianna Edwards and Ron Edwards of Focus Strategic Communications Inc., Oakville, Ontario, Canada, for their considerable efforts in preparing this volume for publication. We would also like to thank the rest of the Focus team for their efforts: Nancy Szostak, designer and formatter; Sarah Waterfield, illustrator; Linda Szostak, copyeditor and proofreader. The authors appreciate the helpful suggestions made by reviewers Sean William Chamberlin and Paul D. Fullagar.

Project Earth Science: Physical Oceanography, Revised 2nd Edition, is published by NSTA Press. We thank everyone at NSTA who helped with this volume, and we especially appreciate the efforts of the publisher, David Beacom. NSTA safety columnist, author, and consultant Ken Roy reviewed the entire manuscript for safety compliance. NSTA Press managing editor Jennifer Horak led NSTA's in-house team for the revised second edition.

Introduction

Project Earth Science: Physical Oceanography is one of the four-volume Project Earth Science series. The other three volumes in the series are *Astronomy, Geology,* and *Meteorology.* Each volume contains a collection of hands-on Activities developed for middle-level students plus a series of Readings intended primarily for teachers, but that could also be useful to interested students.

Additions and Changes to Revised 2nd Edition

The Activities and Readings sections have been rewritten to improve clarity and scientific currency, and to suggest additional teaching and learning strategies. The Resources section at the back of this book has been updated. At the beginning of each Activity, there is now a Planner to quickly provide information about that Activity. Material specifically for students, and material specifically for teachers, is more clearly delineated. There are new sections for students within Activities entitled What Can I Do? and Fast Fact. Additional new sections included for teachers are How Do We Know This?, Safety Alert!, Connections, Differentiated Learning, and Assessment.

Within each Activity, there now is a section for teachers titled Preconceptions. A preconception is an opinion or view that a student might have prior to studying a particular topic. These opinions may not be accurate because the student does not have the correct information or does not understand that information. Each possible preconception that we list with each Activity actually is a misconception. Asking students about their preconceptions at the outset of a new instructional topic can provide useful information about what students already know and what misinformation needs to be corrected so they can have a good understanding of the topic. The preconceptions we list are, of course, only examples of incorrect ideas that some students might have. Most groups of students are imaginative enough to come up with many other preconceptions!

About *Project Earth Science: Physical Oceanography*

This book is divided into three sections: Activities, Readings, and Resources. The Activities in this volume are organized under three broad concepts. First, students investigate the unique properties of water and how these properties shape the ocean and the global environment. Second, students perform Activities investigating the complex systems that lead to the development of currents, waves, and tides. This section focuses on the interactions of wind, water, gravity, and inertia. In the third section, students study the impacts that humans have on the ocean and the marine environment, particularly the effects of pollutants.

An understanding of the concept of density is required for several of the Activities contained in this volume. The Activities are written with the assumption that students have this understanding. The Preconceptions sections are designed to reveal what students actually do understand about density. If their responses indicate a shaky grasp of density, Differentiated Learning offers suggestions for consolidating their knowledge. The series of Activities also provides additional concrete experiences for students to gain mastery in an intuitive way.

At the back of this book, there is also a section on how to construct a wave tank.

A series of overview Readings supports the Activities. By elaborating on concepts presented in the Activities, the Readings are intended to enhance teacher preparation and serve as additional resources for students. The Readings also introduce supplemental topics so that you can link contemporary science to broader subjects and other disciplines.

The Resources provide supplemental materials. The Resources section includes government agencies, organizations, aquaria, coastal reserves, and media. These are annotated and contain the necessary information for gaining access to them.

Creating Scientific Knowledge

Project Earth Science: Physical Oceanography presents a variety of opportunities for you to discuss the creation and evolution of scientific knowledge. For example, students might consider

- how models help develop—yet sometimes restrict—our conceptions of nature
- how scientific knowledge changes over time
- how our choice of measurement scale affects our perceptions of nature and of change

Models and analogies are extremely effective tools in scientific investigation, especially when the subject under study proves to be too large, too small, or too inaccessible for direct study. Although Earth scientists often use models, students must be reminded that models are not perfect representations of the object or phenomenon under study. It is essential that students learn to evaluate models for strengths and weaknesses, such as which phenomena are accurately represented and which are not. When using models, it is good to discuss both their advantages *and* their limitations.

As students learn science, it is easy for them to lose sight of the fact that scientific knowledge evolves. As scientists gather more data, test hypotheses, and develop more sophisticated means of investigation, their understanding of natural phenomena often changes.

With growing information from new technology and expanded understanding, scientific knowledge changes: what seemed impossible to many at the start of the 20th century is accepted in the 21st century. You should emphasize this changing nature of science—it is what makes scientific inquiry special as a form of knowledge—and encourage students to investigate in more detail how scientific knowledge evolves.

Observing and the Problem of Scale

Central to understanding how science evolves is appreciating the limits of our perceptions of change. We observe the world as it *is*, and our thoughts about how it *was* and how it *could be* tend to be quite restricted. That our world is changing constantly can be a difficult concept for students to accept. In several respects, this is a function of the rate at which change sometimes occurs compared to the length of time available to humans for direct observation.

To illustrate this point, ask students to consider the life of an insect that spends its entire existence—from June to August of a single year—in an oak tree. As outside observers, humans can observe seasonal and annual changes in the tree's biology. Due to the relatively short duration of its life, the insect cannot observe these changes.

Likewise, due to the relatively short span of our lifetimes compared to geologic time, people have difficulty appreciating the changes taking place on a million-year scale. The shape of ocean basins change as continents move centimeters per year; average global temperatures may change only a few degrees over thousands of years; sea level changes by millimeters per year. Changes such as these may be imperceptible during a single human life span. It is important for students to understand that while observing these changes may be difficult, Earth is continually changing. Comparing events and changes on different scales can be a difficult concept for students to grasp.

Also, diagrams and models often exaggerate or compress relative sizes to make a certain point more obvious or to make the model small enough to be practical. Sometimes one scale is changed, but others are not. For example, displays of our solar system often accurately depict the relative *distances* between planets but misrepresent planets' relative *sizes*. In oceanography, physical models that are built with an exaggerated vertical scale distort features such as the slope of the continental shelf. It is important for you to discuss the concept of scale and encourage students to raise questions about the various measurement scales used in these Activities.

Getting Ready for Classroom Instruction

The Activities in this volume are designed to be hands-on. In developing them, we tried to use materials that are either readily available in the classroom or inexpensive to purchase. Note that many of the Activities also could be done as demonstrations.

Each Activity has two sections: a Student section and a Teachers' Guide. Each student section begins with Background information to explain briefly, in nontechnical terms, what the Activity is about; the Objective states what students will learn. Then there is Vocabulary, which includes important oceanographic terms students should know. This is followed by a list of the Materials needed and an estimate of the amount of Time that the Activity will take. Following this introduction is a step-by-step Procedure outline and a set of Questions and Conclusions to facilitate student understanding, encourage constructive thinking, and advance the drawing of scientific conclusions.

Each Student section concludes with additional activities for students in What Can I Do? Safety Alert! appears throughout to warn students about dangers in given Activities. Fast Facts, which also appear throughout, are tidbits of information to intrigue students or to provide particulars that will support the Activities. At the end of most Student sections, there are one or more reproducible BLMs (Black Line Masters) for students to fill out.

The Teachers' Guide contains a What Is Happening? section—a more thorough version of the background information given to students. The How Do We Know This? section explains techniques or research methods that oceanographers currently use to generate knowledge related to the Activity. This is followed by a section of possible student Preconceptions, which can be used to initiate classroom discussions. Next comes a summary of What Students Need to Understand, and Time Management discusses the estimated amount of time the Activity will take. The Objective section spells out what students do and learn, while Key Concepts ties the content of the Activity to categories of oceanographic content described on page xii. Preparation and Procedure describes the setup for the Activity. In some cases, we suggest other ways to do the Activity in a section titled Alternative Preparation. Some Activities could be done as a demonstration, for instance, although we advocate giving students the opportunity and responsibility for doing the Activities. Activity 13, on the other hand, involves a large wave tank and is only practical to do as a demonstration.

To challenge students to extend their study of each topic, a section on Extended Learning is provided. For relating the science in each Activity to other disciplines, such as language arts, history, and social sciences, there is a section on Interdisciplinary Study. Connections is a margin feature that links physical oceanography to a similar process or concept in astronomy, geology, or meteorology. The final portion of each Teachers' Guide includes possibilities for Differentiated Learning, Answers to Student Questions, and suggestions for Assessment.

Although the scientific method often is presented as a "cookbook" recipe—state the problem, gather information, form a hypothesis, perform experiments, record and analyze data, and state conclusions—students should be made aware that the scientific method provides an approach to understanding the world around us, an approach that is rarely so straightforward.

For instance, many factors can influence experimental outcomes, measurement precision, and the reliability of results. Such variables must be taken into consideration throughout the course of an investigation.

As students work through the Activities in this volume, make them aware that experimental outcomes can vary and that repetition of trials is important for developing an accurate picture of concepts they are studying. By repeating experimental procedures, students can learn to distinguish between significant and insignificant variations in outcomes. Regardless of how carefully they conduct an experiment, they can never entirely eliminate error. As a matter of course, students should be encouraged to look for ways to eliminate sources of error. However, they also must be made aware of the inherent variation possible in all experimentation.

Finally, controlling variables is important in maintaining the integrity of an experiment. Misleading results and incorrect conclusions often can be traced to experimentation where important variables were not rigorously controlled. You should encourage students to identify experimental controls and consider the relationships between the variables under study and the factors held under control.

Key Concepts

The Activities are organized around three key concepts: the investigation of water and its properties; how the ocean varies spatially and processes that move water vertically and horizontally; and the human impact on Earth's ocean.

Key Concept I: Properties of water
Although water is a common substance, many of its familiar characteristics make it unique among molecules. Its special properties lead to the characteristics of Earth's ocean that make the planet uniquely life-bearing among its neighbors in the solar system. An awareness of the chemical structure and physical characteristics of water underlies an understanding of its importance on Earth.

Key Concept II: Ocean structure and water movement
Movement of water within the ocean occurs through the development of currents, waves, and tides. Deep ocean currents are caused by variations in ocean water density; surface currents result mainly from wind. Waves represent energy in motion and result primarily from the wind as well. Tides are produced by the interaction of forces involving Earth, the Sun, and the Moon.

Key Concept III: Impact of human activities on the ocean
Human activities have an impact on Earth's ocean. The effects are long-lasting and sometimes irreversible.

Project Earth Science: Physical Oceanography and the National Science Education Standards

An organizational matrix for the Activities in *Project Earth Science: Physical Oceanography, Revised 2nd Edition*, appears on pages xvi–xvii. The categories listed along the *x*-axis of the matrix, listed below, correspond to the categories of performing and understanding scientific activity identified as appropriate by the National Research Council's 1996 *National Science Education Standards*.

Subject and Content: Specifies the topic covered by an Activity.

Scientific Inquiry: Identifies the "process of science" (i.e., scientific reasoning, critical thinking, conducting investigations, formulating hypotheses) employed by an Activity.

Unifying Concepts and Processes: Links an Activity's specific subject topic with "the big picture" of scientific ideas (i.e., how data collection techniques inform interpretation and analysis).

Technology: Establishes a connection between the natural and designed worlds.

Personal/Social Perspectives: Locates the specific oceanography topic covered by an Activity within a framework that relates directly to students' lives.

Historical Context: Portrays scientific endeavor as an ongoing human enterprise by linking an Activity's topic with the evolution of its underlying principle.

Project Earth Science: Physical Oceanography hopes to address the need for making science—in this case, physical oceanography—something students do, not something that is done to students. The Standards Organizational Matrix on pages xvi–xvii provides a tool to assist you in realizing this goal.

Safety in the Classroom Practices

The teaching and learning of science today through hands-on, process, and inquiry-based activities make classroom and laboratory experiences effective. Addressing potential safety issues is critical to securing this success. Although total safety cannot be guaranteed, teachers can make science safer by adopting, implementing, and enforcing legal standards and best professional practices in the science classroom and laboratory. Safety in the Classroom Practices includes both basic safety practices and resources. It is designed to help teachers and students become aware of relevant standards and practices that will help make activities safer.

1. When working with glassware, wires, projectiles, or other solid hazards, students should use appropriate personal protective equipment, including safety glasses or goggles, gloves, and aprons.

2. When working with hazardous liquids, indirectly vented chemical splash goggles, gloves, and aprons must be used.

3. Always review Material Safety Data Sheets (MSDSs) with students relative to safety precautions when working with hazardous chemicals.

4. When dealing with hazardous chemicals, an eyewash station within 10-second access is required because of the possibility of a splash accident in the eyes. If there is potential for a body splash, an emergency shower is required within 10-second access.

5. Make sure appropriate laboratory ventilation is used when working with hazardous vapors, fumes, or particulates.

6. Use caution when working with flammables like alcohol. Keep away from flames or sources of sparks. An explosion can occur.

7. When heating liquids other than water, use only heat-resistant glassware (Pyrex- or Kimax-type equipment). Remember that glass labware is never to be placed directly on heating surfaces. Also remember that hot containers are potential hazards. Water may be heated in glassware, but teapots or other types of pans also may be used.

8. When heating liquids on electrical equipment such as hot plates, use ground-fault-protected circuits (GFI). Keep electrical equipment away from water or other liquids—electrical shock hazard.

9. Always remind students of heat and burn hazards when working with heat sources such as hot plates for melting wax, heating water, and more. Remember that it takes time for the hot plate and the objects heated on the hot plate to cool.

10. Lightbulbs can get hot and burn skin. Handle with care.

11. Use caution when working with hot water—it can burn skin.

12. Use caution when working with scissors, wire, or other sharp objects—cut or puncture hazards.

13. If a relatively harmless liquid (e.g., water, dilute chemical) is spilled on the floor, always wipe it up immediately to prevent slip and fall hazards. However, if a spilled liquid (e.g., concentrated acid) is causing, or has the potential to produce toxic fumes, the classroom or lab must be vacated and appropriate emergency authorities called immediately. Teachers must know in advance what to do in this type of emergency.

14. Never consume food or drink that has been either brought into or used in the laboratory.

15. Make sure that all food (e.g., sugar) is disposed of properly so that it does not attract rodents and insects in the lab or classroom.

16. Teachers should always model appropriate techniques before requiring students to cut, puncture, or dissect, and so on.

17. Wash hands with soap and water after doing activities dealing with hazardous chemicals.

18. Markers can have volatile organic compounds (VOCs) that can irritate the eyes, nose, and throat. Use in well-ventilated areas, or use only low-VOC markers.

19. Clear the area of falling or tripping hazards such as desks, other furniture, or equipment before conducting activities that require open floor space (e.g., Activity 11).

20. Know the source for any materials used in this Activity. Never use garbage or other refuse. Also check for mold or fungi. Some students are allergic to these organisms.

For additional safety regulations and best professional practices, go to
NSTA: Safety in the Science Classroom: *www.nsta.org/pdfs/SafetyInTheScience Classroom.pdf*

NSTA Safety Portal: *www.nsta.org/portals/ safety.aspx*

XV

Standards Organizational Matrix

Activity	Subject and Content	Scientific Inquiry	Unifying Concepts and Processes
Activity 1 **A Pile of Water**	Surface tension and cohesion of water	Exploring, estimating, and predicting	Predicting based on an ordered universe
Activity 2 **A Sticky Molecule**	Structure of water molecules, and ensuing polarity	Modeling, visualizing	Structure and properties of matter
Activity 3 **Over and Under—** **Why Water's Weird**	Density of solid versus liquid states of matter	Observing	Structure and properties of matter
Activity 4 **How Water Holds Heat**	Specific heat of sand and water	Measuring and graphing data	Measuring relative rate of change
Activity 5 **Water—The Universal Solvent**	Solubility	Observing and recording data in an organized way	Structure and properties of matter
Activity 6 **Won't You BB My Hydrometer?**	Measuring density of freshwater and salt water	Measuring	Measuring and instrumentation
Activity 7 **Ocean Layers**	Salinity/density layers in water	Modeling and observing	Change within open systems
Activity 8 **The Myth of Davy Jones's Locker**	Density layering of ocean	Modeling and observing	Effect of density on ocean's vertical structure
Activity 9 **Estuaries—Where the Rivers** **Meet the Sea**	Mixing of estuaries	Modeling and observing	Modeling change within estuaries
Activity 10 **Current Events in the Ocean**	Surface ocean currents	Modeling and observing	Change due to atmosphere/ocean interactions
Activity 11 **Body Waves**	Wave motion and energy	Modeling and observing	Energy flow via waves
Activity 12 **Waves and Wind in a Box**	Wind's influence on waves	Modeling, observing, and experimenting	Energy flow in open systems
Activity 13 **Tanks a Lot—Activities for** **a Wave Tank (Teacher** **Demonstration)**	Modeling wave formation, properties, and behavior	Modeling, observing, measuring, and experimenting	Energy flow in open systems
Activity 14 **Plotting Tidal Curves**	Tidal patterns	Graphing, analyzing data	Patterns of change within systems
Activity 15 **Tides Mobile**	Why tides form	Modeling and explaining	Modeling to explain patterns of change within systems
Activity 16 **The Bulge on the Other Side** **of Earth**	Inertia as a factor in tides	Modeling and explaining	Modeling to explain patterns of change within systems
Activity 17 **Oily Spills**	Oil spills and their mitigation	Modeling, experimenting, observing, and analyzing	Modeling to explore systems
Activity 18 **Forever Trash**	Decomposition of ocean pollutants	Experimenting, observing, and analyzing	Experimenting with conditions to explore changes in a system

Technology	Personal/Social Perspectives	Historical Context	Key Concept
			I
			I
			I
			I
			I
Building a hydrometer	Improvements in measuring devices allow science to advance		I
			II
	Science in society	Seafaring mythology	I, II
	Influence of human activity on estuaries		II, III
			II
			II
Building a wave tank			II
Measuring waves using a wave tank			II
	Waves as a natural hazard	Historic wave events	II
Building a planet/tide model			II
Building an inertia/tide model			II
Testing technology to clean up spills	Risks and benefits of using oil	Historic oil spills	III
	Risks of designing synthetic materials		III

Activities at a Glance Matrix

Activity	Pages	Subject and Content	Objective	Materials
Activity 1 A Pile of Water	1–11	Surface tension and cohesion of water	Investigate a specific property of water: its ability to "stick" to itself.	The teacher will need (for preparation ahead of Activity): food coloring, water Each group (of five groups) will need: one of five containers varying in shape and size of opening, several rolls of pennies Each student will need: penny, eyedropper, small beaker or clear plastic cup (may be shared), paper towel, other assorted coins (quarters, nickels, etc.)
Activity 2 A Sticky Molecule	13–21	Structure of water molecules and ensuing polarity	Construct a model of the water molecule to explore the concepts of polarity and hydrogen bonding.	Each student will need: paper molecule pattern, scissors, glue, crayons or markers (red and blue), blank paper
Activity 3 Over and Under— Why Water's Weird	23–31	Density of solid versus liquid states of matter	Observe the behavior of different substances, including water, in their solid and liquid states.	Each group will need: several cups of ice cubes, several cups of unflavored gelatin cubes—clear and colored, vegetable shortening, teaspoon, spoon or tongs, hot plate with wire gauze screen, three 250 ml beakers, indirectly vented chemical splash goggles, gloves, aprons
Activity 4 How Water Holds Heat	33–43	Specific heat of sand and water	Compare the specific heat of sand and water.	Each group will need: two thermometers (nonmercury), ring stand, two utility clamps, two 240 ml (8 oz.) polystyrene foam cups, sand, water, string, lamp with reflector and 200-watt bulb, clock or watch with a second hand, balance, graduated cylinder, ruler or meter stick, GFI-protected circuit
Activity 5 Water— The Universal Solvent	45–55	Solubility	Explore the solubility of various substances in water as compared with other liquids.	Each group will need: water, mineral oil (or baby oil), isopropyl alcohol (70%), Epsom salts (Mg_2SO_4), baking soda ($NaHCO_3$), table salt (NaCl), granular sugar, red wax marking pencil, five test tubes with rubber stoppers, small spoon (1/8 teaspoon will work), test tube rack, graduated cylinder, black construction paper, indirectly vented chemical splash goggles, gloves, aprons, MSDSs for all hazardous materials

Time	Vocabulary	Key Concepts	Margin Features
50 minutes	Atom, Molecule, Molecular structure	I	Safety Alert!, Fast Fact, What Can I Do?, Connections, Resource
50 minutes	Molecule, Atom, Chemical bonds	I	Safety Alert!, Fast Fact, What Can I Do?, Connections, Resources
30–50 minutes	Density	I	Safety Alert!, Fast Fact, What Can I Do?, Connections, Resources
50 minutes	Specific heat, Calorie	I	Safety Alert!, Fast Fact, What Can I Do?, Connections, Resources
50 minutes	Solvent, Solubility	I	Safety Alert!, Fast Fact, What Can I Do?, Connections, Resources

Activity	Pages	Subject and Content	Objective	Materials
Activity 6 Won't You BB My Hydrometer?	57–69	Measuring density of freshwater and salt water	Understand the implications of water density by building and using a hydrometer to measure the densities of freshwater and saltwater samples.	For Part 1, each group will need: plastic transfer pipette, sharp scissors, fine-tip permanent marking pen, metric ruler, 20 BBs, 500 ml beaker, masking tape, modeling clay, food coloring, pickling salt, waste container, towels or rags for cleanup, teaspoon (5 ml) or tablespoon (15 ml) For Part 2, each group will need: large jar or beaker, pickling salt, soupspoon (large enough to hold an egg), 100 ml graduated cylinder, hard-boiled egg, ruler or straightedge, 5 ml metric measuring spoon (a teaspoon), hydrometer (from Part 1)
Activity 7 Ocean Layers	71–79	Salinity/density layers in water	Investigate what happens when ocean water, brackish water, and river water contact one another.	Each group will need: cafeteria tray, slice of clay 3 cm thick, clear plastic straw (about 10 cm long), three 250 ml clear plastic cups containing 25 ml each of the colored solutions, 250 ml clear plastic cup (waste container for used solutions), three medicine droppers or plastic pipettes, one or two sheets of white paper, towels or rags for cleanup
Activity 8 The Myth of Davy Jones's Locker	81–91	Density layering of ocean	Investigate some of the properties of water that could explain the myth of Davy Jones's Locker.	For the demonstration, the teacher will need: three buckets (4 L or 1 gal. size), eight small screw-top vials (or eight test tubes and stoppers), plastic cylinder with an end cap (122 cm [4 ft.] tall and 4 cm [1.5 in.] in diameter), ring stand with clamp to hold cylinder, large funnel, 1 to 2 m length of rubber tubing with U-shaped glass tubing in one end, package of BBs, hot plate and pan or coffee heater, pickling salt (produces a clear brine), ice, water, red wax marking pencil, optional: food coloring For the Activity, each student or group will need: three 600 ml beakers, eight small screw-top vials (or eight test tubes and stoppers), plastic cylinder with an end cap (122 cm [4 ft.] tall and 4 cm [1.5 in.] in diameter), ring stand with clamp to hold cylinder, large funnel, 0.5 m length of rubber tubing with U-shaped glass tubing in one end, package of BBs, pickling salt, water, red wax marking pencil

Time	Vocabulary	Key Concepts	Margin Features
90–100 minutes (45–50 minutes for each part)	Hydrometer, Density	I	Safety Alert!, Fast Fact, What Can I Do?, Connections, Resources
50 minutes	Estuary, Brackish water	II	Safety Alert!, Fast Fact, What Can I Do?, Connections, Resources
Demonstration: 25–30 minutes Activity: 50 minutes		I, II	Safety Alert!, Fast Fact, What Can I Do?, Connections

Activity	Pages	Subject and Content	Objective	Materials
Activity 9 **Estuaries— Where the Rivers Meet the Sea**	93–101	Mixing of estuaries	Investigate how water mixes in estuaries.	Each group of students will need: clear Pyrex glass loaf pans, 500 ml "ocean water" (see Preparation section), 500 ml "river water" (see Preparation section), pickling salt, blue and yellow food coloring, graduated cylinder, pencil or pen, 20 cm masking tape, spoon, waste container for used solutions, towels or rags for cleanup
Activity 10 **Current Events in the Ocean**	103–115	Surface ocean currents	Model how landforms and wind affect ocean surface currents.	Each group of three to four students will need: indirectly vented chemical splash goggles and aprons; baking pan, 30 cm × 45 cm × 3 cm (12 in. × 18 in. × 1.5 in.) deep, painted black inside; white chalk; modeling clay; colored pencils; a plastic drinking straw with a flexible elbow for each student; black permanent marker (no or low VOC); 400 ml rheoscopic fluid; towels or rags for cleanup
Activity 11 **Body Waves**	117–125	Wave motion and energy	Investigate the energy of a wave and the motion of the medium through which a wave travels.	None
Activity 12 **Waves and Wind in a Box**	127–135	Wind's influence on waves	Investigate the relationship between wind and waves.	Each group will need: one or two large plastic trash bags (preferably white), 2 kg sand (optional), two sturdy cardboard boxes (75 cm × 28 cm × 5 cm or comparable size), two-speed fan or hair dryer, scissors, packing or duct tape, stopwatch or watch with second hand
Activity 13 **Tanks a Lot— Activities for a Wave Tank (Teacher Demonstration)**	137–147	Modeling wave formation, properties, and behavior	A: Teach students about selected wave characteristics and properties. B: Show the effect of water depth on wave speed. C: Show the orbital motion of particles in an ocean wave, and two aspects of wave–coast interactions: breakers and sand transport.	For demonstration, teacher will need: water, wave tank, stopwatch, pebbles, floating objects (e.g., Ping-Pong balls, eyedroppers, corks)
Activity 14 **Plotting Tidal Curves**	149–159	Tidal patterns	Plot tide data for a period of one month and draw the tidal curve for this data.	Each student will need: pencil with eraser, red pen or a bright color crayon, scissors, clear tape, ruler

Time	Vocabulary	Key Concepts	Margin Features
Preparation 25–30 minutes Activity 50 minutes		II, III	Safety Alert!, Fast Fact, What Can I Do?, Connections, Resources
50 minutes	Ocean currents, Gulf Stream, California Current, Rheoscopic fluid	II	Safety Alert!, Fast Fact, What Can I Do?, Connections, Resources
30 minutes or less	Oscillate	II	Safety Alert!, Fast Fact, What Can I Do?, Connections, Resources
100 minutes		II	Safety Alert!, Fast Fact, What Can I Do?, Connections, Resources
50 minutes, more time optional	Crest, Trough	II	Safety Alert!, Fast Fact, Connections, Resources
100 minutes	Bathymetry, Global forcing	II	Fast Fact, What Can I Do?, Connections, Resources

Activity	Pages	Subject and Content	Objective	Materials
Activity 15 Tides Mobile	161–169	Why tides form	Construct a mobile that shows the relationship among the Sun, Moon, and Earth, and use this mobile to investigate how tides are created.	Each group will need: scissors, coat hanger, string, meter stick or dowel, modeling clay, tape, yellow construction paper, pencil, paper clip
Activity 16 The Bulge on the Other Side of Earth	171–181	Inertia as a factor in tides	Demonstrate how the rotation of the Earth–Moon system accounts for the bulge of water on the side of Earth facing away from the Moon.	Each group will need: safety glasses or goggles for each student, Styrofoam ball (15 cm diameter), Styrofoam ball (6 cm diameter), string, dowel (1 m long and 0.5 cm diameter), dowel (0.5 m long and 0.5 cm diameter), masking tape, two weights (about 15 g each), 473 ml (16 oz.) drink bottle, ruler
Activity 17 Oily Spills	183–191	Oil spills and their mitigation	Explore the effectiveness of different methods for cleaning up oil spills.	Each group of four or more students will need: dish pan or plastic tub, tap water, vegetable oil or heavy olive oil (500 ml), cotton string (approximately 1 m long), several drinking straws (cut in half), paper towels, several small pieces of polystyrene foam (such as packing material), 10 ml liquid detergent, sand (125 g), diatomaceous earth (125 g), feather
Activity 18 Forever Trash	193–204	Decomposition of ocean pollutants	Observe the breakdown of various materials in water and in sand.	Each group will need: small piece of paper (10 cm × 10 cm); 10 cm × 10 cm scraps of cloth (cotton, rayon, wool, polyester, nylon, etc.); small sheets of aluminum foil, waxed paper, plastic wrap; aluminum soda can tabs; plastic bag (sandwich size); pieces of a plastic grocery bag; plastic six-pack holder; plastic bottle cap; hard candy in a plastic wrapper; unwrapped hard candy; rubber balloon; polystyrene foam packing peanuts; starch-based packing peanuts; sand containing organic matter; shoe box or small individual containers; beaker; salt (for salt water)

Time	Vocabulary	Key Concepts	Margin Features
50 minutes	Gravitational forces, spring tides, neap tides	II	Safety Alert!, Fast Fact, What Can I Do?, Connections, Resources
50 minutes	Inertia, Axis of rotation, Center of gravity	II	Safety Alert!, Fast Fact, What Can I Do?, Connections, Resources
50 minutes	Dispersant	III	Safety Alert!, Fast Fact, What Can I Do?, Connections, Resources
50 minutes twice, with at least a week between	Biodegradable	III	Safety Alert!, Fast Fact, What Can I Do?, Connections, Resources

Activity 1 Planner

Activity 1 Summary

Students count the number of drops of water they can add onto coins before the water spills over. They then apply what they learn by predicting and then testing the number of coins they can add to different containers filled with water before the water overflows. These exercises teach students about the ability of water to "stick" to itself.

Activity	Subject and Content	Objective	Materials
A Pile of Water	Surface tension and cohesion of water	Investigate a specific property of water: its ability to "stick" to itself.	The teacher will need (for preparation ahead of Activity): food coloring, water Each group will need (class will be divided into five groups): one of five containers varying in shape and size of opening, several rolls of pennies Each student will need: penny, eyedropper, small beaker or clear plastic cup (may be shared), paper towel, other assorted coins (quarters, nickels, etc.)

Time	Vocabulary	Key Concept	Margin Features
50 minutes	Atom, Molecule, Molecular structure	I: Properties of water	Safety Alert!, Fast Fact, What Can I Do?, Connections, Resource

Scientific Inquiry	Unifying Concepts and Processes
Exploring, estimating, and predicting	Predicting based on an ordered universe

A Pile of Water

Background

We observe and use water every day. It makes life on Earth possible. Water covers nearly three-fourths of Earth's surface and affects almost all living and nonliving things. Because it is so abundant, it may not seem unusual, but water is unique when compared to other substances in the universe. In fact, its properties are quite different from those of other substances even here on Earth. For instance, it is the only substance on Earth that occurs naturally in all three states—solid (e.g., an iceberg), liquid (e.g., ocean water), and gas (e.g., steam or vapor in a cloud). (See **Figure 1.1**.)

Most substances (water, air, dirt, etc.) are made up of **atoms**. Atoms are arranged in a specific way, often forming a **molecule**. The makeup of a water molecule—or any molecule—is called **molecular structure**. A substance's molecular structure is responsible for its properties and governs how it interacts with other things on Earth. This Activity introduces and explores one specific property of liquid water.

Vocabulary

Atom: The basic unit of matter.

Molecule: The most basic unit of many substances; it has a specific arrangement of atoms.

Molecular structure: The arrangement of atoms in a specific molecule.

Objective

Investigate a specific property of water: its ability to "stick" to itself.

Topic: water cycle
Go to: *www.scilinks.org*
Code: PESO 001

Activity 1

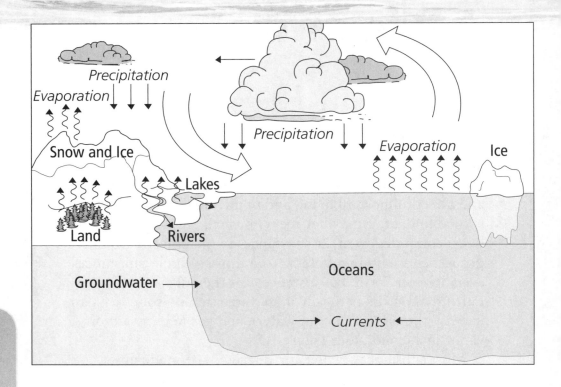

Materials

Each group will need (class will be divided into five groups):

- one of five containers varying in shape and size of opening
- several rolls of pennies

Each student will need
- penny
- eyedropper
- small beaker or clear plastic cup (may be shared)
- paper towel
- other assorted coins (quarters, nickels, etc.)

Time

50 minutes

SAFETY ALERT

1. Be careful to quickly wipe up any spilled water on the floor—slip and fall hazard.

2. Wash hands with soap and water upon completing the lab.

Procedure

Part 1

1. Explore the behavior of water with a single coin by placing a penny on a piece of paper towel.

2. Estimate the number of water drops you can pile on the penny before the water runs over its edge. Record your estimate in the table on **BLM 1.1**.

3. Place water on the penny drop by drop. Working with the other members of your group, develop a technique that allows you to put the most drops on your penny. You may want to put the drops on in different areas of the penny or from different heights. Count each drop until the water spills over. Record your results in the table on **BLM 1.1**.

4. Make a sketch of the water on the surface of the penny just before the water spilled over.

5. Based on what you observed with the penny, make a prediction comparing the number of drops you could pile on a nickel, dime, or quarter. Remember that the area of the different-size coins is important to your predictions. Record these predictions in the table on **BLM 1.1**.

6. Repeat steps 2, 3, and 4 with each coin to test your hypothesis.

7. Compare the techniques you used that allowed you to put the most drops on a coin.

Part 2

1. Your teacher has filled five different containers with colored water. (See **Figure 1.2**.) Observe the containers closely. Record brief descriptions of the size, shape, and other characteristics of the containers that help you distinguish among them.

2. Predict which of the five containers will hold the greatest number of pennies without spilling over. Record your prediction in the table on **BLM 1.2**.

3. Estimate the number of pennies that would have to be added to each container to make it spill over. Record your estimates in the table on **BLM 1.2**. You will have a chance to revise these estimates in step 5 below.

4. You will add pennies to a container, one at a time, counting how many you add before water spills over the lip. Add them this way:

   (a) Hold the penny so its edge will enter the water first (not flat).
   (b) Hold the penny over the center of the container opening, no more than 5 cm above the surface of the water.
   (c) Release the penny and let it drop into the water.

5. Based on what you learned by adding pennies to one container, revise your predictions about how many pennies the other container can hold before overflowing. Record your predictions in the table on **BLM 1.3**.

Questions and Conclusions

1. Describe the way water "sits" on the penny.

2. Why do some pennies hold more water droplets than others?

3. Why do you think water piles up on the penny, rather than spilling over the edges immediately?

4. Suggest reasons why the five containers hold a different number of pennies before spilling over.

5. In Part 2, step 5, did you change your predictions from step 3 before testing your original prediction? Describe how this is consistent with the way scientists test their ideas.

Fast Fact

One milliliter of water has 3.34×10^{22} molecules in it. This is the shorthand way of writing a long number with many zeros: 3.34×10^{22} means that the decimal point in 3.34 actually goes 22 places to the right, so that the long version of the number is 33,400,000,000,000,000,000,000!

Figure 1.2
Five possible types of containers to be used in Activity 1

What Can I Do?

You could figure out how many milliliters of water, on average, you put on a penny, and work out how many molecules that is. (How could you determine how many milliliters are in each drop of water?)

Data Table: Predicted and Observed Results

Item	Number of Drops	
	Prediction	Observation
Penny		
Nickel		
Dime		
Quarter		

Data Table: Initial Predictions

Container Predicted to Hold the Most Pennies Before Overflowing:		
Container Number	Description	Number of Pennies Predicted to Cause Overflow
1		
2		
3		
4		
5		

Data Table: Revised Predictions and Observed Results		
Container Predicted to Hold the Most Pennies Before Overflowing:		
Container Number	Number of Pennies Predicted to Cause Overflow	Number of Pennies Required to Cause Overflow
1		
2		
3		
4		
5		

A Pile of Water

What Is Happening?

The properties of water play an integral role in the development and maintenance of Earth's environment and its ability to sustain life. Although water is a unique substance in the universe, it is so common on Earth that many students may expect other substances to have similar properties.

Water exhibits characteristics that are unusual. For instance, solid water floats in liquid water (unlike most solids, which are denser than their liquid and therefore sink); large amounts of energy must be added to water to achieve relatively small changes in temperature (heat capacity); and water molecules tend to "stick" to each other (cohesion) and to other molecules (adhesion). Later Activities will explore the first two properties, while this Activity introduces students to liquid water's ability to "stick" to itself.

Water molecules "stick," or are attracted to one another, because water has an uneven distribution of electrical charge. (It is a "polar molecule.") Each molecule has a positive end, or "pole," and a negative pole. The positive end of one molecule and the negative end of another molecule attract each other. This attraction, called hydrogen bonding, is strong enough to hold water molecules together. The force of hydrogen bonds causes water to fall in drops and to dome up on flat surfaces or containers full of water.

When placed on coins, the molecules of water form flexible piles that stay together because of hydrogen bonding. This phenomenon—of water "piling up"—is due to surface tension. Liquid water has an extremely high surface tension because of its molecular structure and the hydrogen bonding between molecules.

Objective

Investigate a specific property of water: its ability to "stick" to itself.

Key Concept

I: Properties of water

Materials

The teacher will need (for preparation ahead of Activity):
- food coloring
- water

Each group will need (class will be divided into five groups):
- one of five containers varying in shape and size of opening
- several rolls of pennies

Each student will need
- penny
- eyedropper
- small beaker or clear plastic cup (may be shared)
- paper towel
- other assorted coins (quarters, nickels, etc.)

Time

50 minutes

How Do We Know This?

How do we measure surface tension in liquids?

The traditional way to measure tension in liquids is by slowly pulling out a ring (called a Du Nouy ring, often made of platinum) from the surface of the liquid. We measure the force required to raise the ring from the liquid's surface; this force is a direct measure of the surface tension. Another method uses a vertical plate (known as a Wilhelmy plate) and a precision balance to measure the force caused by placing the plate in contact with the liquid surface.

Preconceptions

Explain to students, "Physical properties are descriptions of a substance by itself, like its boiling point and melting point. Chemical properties are descriptions of how a substance reacts in the presence of another, like the way iron rusts in the presence of oxygen." Then ask students, "What do you think you know about physical and chemical properties of water?" You could ask them as a Think-Pair-Share, a journal entry, a concept map, or a class discussion. You can also ask if students have watched insects walking on water. What did they see? Did they have any questions about it? Some preconceptions that students may have are as follows:

- What is the shape of a raindrop? Most people think raindrops have a teardrop shape when in reality they are spherical due to surface tension.
- Because water is such a part of our everyday lives, it is typical of compounds.
- All liquids mound the same way that water does on a penny.

What Students Need to Understand

- Water is unique among substances.
- The high surface tension of water, which results in water "piling up" on a flat surface, is just one of the unusual properties of water. Water's characteristics are important in determining how water interacts with other substances.
- The characteristics of water make life possible on Earth. One of these characteristics, surface tension, is investigated in this Activity. It allows water striders to "walk" on the surface of water. It also aids in capillary action, which allows groundwater to move through soil, wells to function, and plants to transport water from their roots upward.

Time Management

Students can do both parts of this Activity in 50 minutes. Be sure to leave at least 5 minutes at the end of class for them to compare their predictions for Part 2 to their classmates' results.

Preparation and Procedure

Choose five containers, each differing in volume and mouth shape and size. The containers may be of like or different materials (plastic, glass, etc.). The material used affects the surface tension to a much smaller degree than the diameter of the opening. Possible containers might include an apple juice jar, a wide-mouthed plastic cup, a petri dish, a two-liter plastic drink bottle, and a canning jar. It is important that the containers chosen for the Activity have a variety of opening sizes, from small to large. *The success of the Activity depends on using a wide variety of containers*. All containers should be transparent.

Fill each container with water and add a different color of food coloring to each. Be sure all containers are filled to the point where the water is exactly level with the opening of the container. The water should not form a depression (meniscus) or be domed up in the center of the opening. You might want to do this part of the Activity ahead of time. Also, check the water level after each group has used a container.

Give students an opportunity to record their results on the board and make comparisons with other groups in the class. Ask students to compute a class average for the results from the table on **BLM 1.1**, then construct two bar charts: one tallying students' predictions, and the other tallying the actual number of drops that fit on each coin. What conclusions emerge from looking at the data in this way? (Did most students predict too many drops or too few?)

Extended Learning

- Challenge students to float a paper clip on a petri dish full of water. With the paper clip floating, add a drop or two of liquid detergent or soap. Watch what happens and ask students to try to refloat the paper clip. Soap reduces the surface tension of the water, making the paper clip sink. (Explain this *after* students have watched the paper clip sink).

- Compare the surface tension of tap water and salt water. Although the addition of impurities—like salt—decreases the cohesion between water molecules, it also increases the density of water (density will be explored in Activity 3). For example, the presence of large quantities of salt allows objects that would sink in freshwater to float on the surface of water in the Dead Sea and in the Great Salt Lake. This may be confusing to students. The ability to float is the result of a difference in density, rather than an increase in surface tension.

- Ask students why water forms drops. What factors affect the size of a water drop? How would that compare with drops from other liquids?

- Challenge students to repeat Part 2 using liquids other than water. Their predictions before testing will expose any misconceptions that all liquids behave as water does. Rubbing alcohol, white vinegar, and vegetable oil make good alternatives.

- Students should not try to experiment with mercury, but they can read about the differences between water and mercury. Which one has more surface tension? What are the effects on the formation of drops? What shape (concave or convex) does the surface of each liquid have when in a glass container?

Interdisciplinary Study

Explore either capillary action, the mechanism by which groundwater moves and trees and plants transport water from their roots throughout the plant, or "water striders"—animals that take advantage of water's surface tension to live on the surface of the water.

Differentiated Learning

This is a concrete Activity with which many students will be successful. Ask students who are skilled in mathematics or who enjoy math challenges to work out the number of molecules they added to the pennies using scientific notation, as described in How Do You Know This?

Answers to Student Questions

1. The water forms a dome on the surface of the penny. (Note: The hydrogen bonds hold the water molecules together, allowing many more drops than would be expected to pile on the penny.)

2. Pennies hold different numbers of drops because they are not all exactly the same. Some pennies have worn edges, while others are dirty or dented. There may also be a difference between the head and tail sides of the pennies. (Note: There also may be differences in the size of the drops. Encourage students to develop techniques that "standardize" their drop size.)

3. Answers will vary since students have not yet been introduced to hydrogen bonding. Students may create explanations, though, that are fairly accurate. Encourage students to formulate hypotheses explaining why the water piled up. Address these hypotheses through discussion. The answer: The hydrogen bonding between water molecules holds the molecules together. When the number of water molecules gets too large for the cohesive forces to hold together, the water spills over the side.

4. Again, answers will vary. Ask students to formulate and discuss their hypotheses. The openings or mouths of the containers differ in surface area and shape. Containers with large round openings will hold a greater number of pennies than those with small round openings. There may also be differences in the adhesive forces between water and the material from which the containers are made.

5. Answers will vary. (Note: Most students greatly underestimate the number of pennies it takes to make the water spill over the sides of the containers. Therefore, after observing the water dome up on the coins, most students will increase their estimates.) Scientists often form their initial ideas based on intuition or initial observations. Hypotheses are revised as additional data are collected through experimentation or further observations.

Assessment

- Ask students to revisit their responses to your questions in Preconceptions. What have they learned about the properties of water? How would they respond now?
- You can also grade student answers to questions.

Resource

www.haydenplanetarium.org/
tyson/read/1997/03/01/on-
being-round

Activity 2 Planner

Activity 2 Summary

Students use paper models of hydrogen and oxygen nuclei and electrons to explore the polar nature of water and of bonding within a molecule and among molecules.

Activity	Subject and Content	Objective	Materials
A Sticky Molecule	Structure of water molecules and ensuing polarity	Construct a model of the water molecule to explore the concepts of polarity and hydrogen bonding.	Each student will need: paper molecule pattern, scissors, glue, crayons or markers (red and blue), blank paper

Time	Vocabulary	Key Concept	Margin Features
50 minutes	Molecule, Atom, Chemical bonds	I: Properties of Water	Safety Alert!, Fast Fact, What Can I Do?, Connections, Resources

Scientific Inquiry	Unifying Concepts and Processes
Modeling, visualizing	Structure and properties of matter

A Sticky Molecule

Activity 2

Background

Water is one of the simplest chemical substances on Earth, and yet we must have it to live. A person can survive only about a week without water. Water is so common that we often ignore some of its characteristics. Pure water is clear, with no odor or taste. These are three obvious properties of water; others are less obvious. For example, compared to other liquids, it takes a lot of heat to make water hot and even more to make it boil.

Knowing the molecular structure of water—or any other substance—will help you understand many of its properties. In every water **molecule**, two hydrogen **atoms** are joined to one oxygen atom by forces called **chemical bonds**. The chemical bond between hydrogen and oxygen occurs when each of the two hydrogen atoms shares electrons with the oxygen atom. This is shown in **Figure 2.1**.

In this Activity, you will learn how hydrogen and oxygen join and investigate some characteristics of the bond between them. This will help you understand the properties of water. You can explore less obvious properties of water in the next several Activities.

Vocabulary

Molecule: The most basic unit of many substances; it has a specific arrangement of atoms.

Atom: The basic unit of matter.

Chemical bonds: The forces between atoms that hold a molecule together. There are different types.

Objective

Construct a model of the water molecule to explore the concepts of polarity and hydrogen bonding.

Topic: water properties
Go to: *www.scilinks.org*
Code: PESO 002

Activity 2

Figure 2.1
The bond between hydrogen and oxygen occurs when they share two electrons— one from hydrogen and one from oxygen. In the water molecule, we have two hydrogen atoms and one oxygen atom joined by sharing electrons.

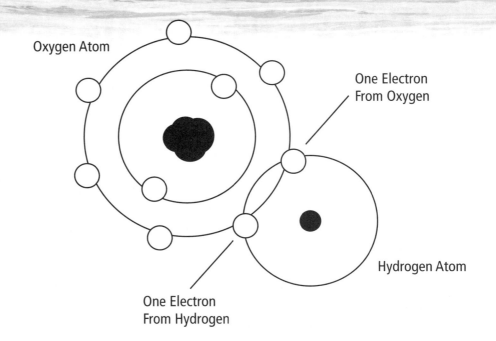

Oxygen Atom

One Electron From Oxygen

Hydrogen Atom

One Electron From Hydrogen

Materials

Each student will need
• paper molecule pattern
• scissors
• glue
• crayons or markers (red and blue)
• blank paper

Time

50 minutes

Procedure

1. Locate the Water Molecule Pattern Sheet (**BLM 2.1**). Color both hydrogen atoms and nuclei blue, and the oxygen atom and nucleus red, leaving the electrons uncolored. Cut out all the pieces (atoms, nuclei, and electrons) of the water molecule.

2. Remember that in every water molecule, two hydrogen atoms are bonded to one oxygen atom. Before gluing the hydrogen atoms to the oxygen atom, try different arrangements of the atoms; use a blank piece of paper to sketch out at least three different ways that the two hydrogen atoms and one oxygen atom could be joined in a single molecule.

3. In the 20th century, scientists discovered the actual shape of a water molecule. The hydrogen atoms are attached to the oxygen atom in a way that makes the molecule look like the head and ears of a mouse viewed head-on. The oxygen atom is the head of the mouse, and the two hydrogen atoms are its ears.

4. Based on this description, glue the hydrogen and oxygen atoms together. The glue represents the bonding between the hydrogen atoms and the oxygen atom.

5. Glue a nucleus in the center of each atom so that the nucleus covers the letter representing the atom ("O" or "H").

6. Count the electrons you have cut out. You should have 10 electrons.

    (a) Glue two of these electrons to the oxygen atom, placing them on opposite sides of the dashed circle.

(b) Remember that each bond between hydrogen and oxygen atoms forms by sharing two electrons. (See **Figure 2.1.**) At each place where the edge of the hydrogen atom crosses the edge of the oxygen atom, glue two electrons.

(c) Glue the rest of the electrons (you should have four left) to the oxygen atom, spacing them evenly around the solid outer circle.

7. Because of the way hydrogen and oxygen bond, each hydrogen atom has a slightly positive charge, and the oxygen atom has a slightly negative charge. Draw a minus ("–") sign on the oxygen atom and a plus ("+") sign on each hydrogen atom. Do you notice that the "+" signs are on one end of the molecule and the "–" sign is on the other end? This gives the molecule opposite charges on either end—similar to a bar magnet's north and south poles. Molecules with their "+" and "–" charges arranged like this are called polar molecules. Water is considered a strongly polar molecule.

Since water molecules are polar, they "stick" together much like magnets (see questions 5 and 6). This sticking together is one reason for water's unusual properties, such as surface tension, which you might have explored in Activity 1.

Questions and Conclusions

1. Name the elements found in a water molecule.

2. What is the ratio of hydrogen atoms to oxygen atoms in a water molecule?

3. A molecule is a combination of atoms that are bonded together. How are the oxygen and hydrogen atoms of a water molecule held together?

4. Describe a polar molecule.

5. If one object has a positive charge and one has a negative charge, what will they tend to do to each other? (Hint: Have you ever heard the saying, "opposites attract"?)

6. Using your answer to question 5, why do you think water molecules tend to "stick" together? What are some different ways that two water molecules could be positioned so that they stick together?

7. Note the different orientation of the "mouse heads" in **Figure 2.1.** How can you explain this?

Fast Fact

Water is abundant on Earth's surface (that is why it looks like a blue marble from the Moon), but freshwater is not. Less than 4% of water is fresh, and less than 0.007% is in lakes and rivers, the source of most of our water supply.

What Can I Do?

Determine how much water you use in your everyday shower. Run the shower into a bucket (note the volume of the bucket), and time how long it takes to fill it. Time your showers for a week or so to get a reasonable average time, and then work out how much water you typically use (and heat). Can you reduce the volume of water you use?

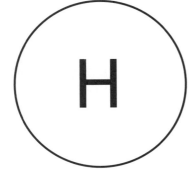

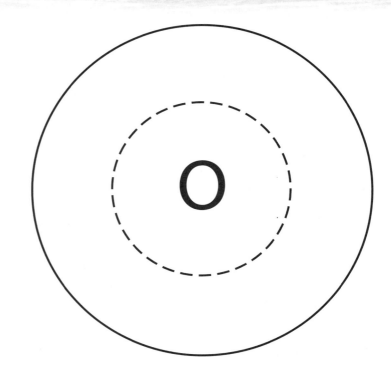

Hydrogen
Nucleus
1 proton

Oxygen
Nucleus
8 protons
8 neutrons

Hydrogen
Nucleus
1 proton

Electrons

A Sticky Molecule

What Is Happening?

Water is a simple molecule, yet no other substance may be more important for life as we know it. A water molecule is made up of only two elements—hydrogen and oxygen. The chemical formula of water is H_2O, meaning that in every water molecule, two hydrogen atoms are bonded to one oxygen atom. Exploring the nature of the bond between hydrogen and oxygen, and observing the resulting shape of the water molecule, will help students gain insight into water's unique properties.

The atoms in a water molecule bond to each other by sharing electrons, called a covalent bond. The hydrogen and oxygen atoms bond by sharing between them hydrogen's lone electron—a hydrogen atom has only one electron—and one electron from oxygen. The sharing, however, is not equal. Oxygen's pull on hydrogen's electron is slightly stronger than hydrogen's own pull. Therefore, the electron spends most of its time closer to oxygen than to hydrogen. Because electrons carry a negative charge, the presence of the extra electron results in a partial negative charge on the oxygen atom. The absence of the electron from hydrogen results in a partial positive charge there. The result is that a molecule of water has poles, like a magnet.

This bond, which creates a partial negative charge on the one oxygen atom and partial positive charges on the two hydrogen atoms, is called a polar covalent bond. The resulting molecule is called a polar molecule.

Objective

Construct a model of the water molecule to explore the concepts of polarity and hydrogen bonding.

Key Concept

I: Properties of water

Materials

Each student will need

- paper molecule pattern
- scissors
- glue
- crayons or markers (red and blue)
- blank paper

Time

50 minutes

How Do We Know This?

How do we know the shape of water molecules? How do we measure the angles between atoms?

The two basic techniques to determine the shape of molecules use spectroscopic and nuclear magnetic resonance (NMR) methods. In spectroscopy, infrared or microwave radiation can provide information about the molecule geometry from vibrational (changing the vibration state of the electrons usually with infrared radiation) and rotational (changing the spin of the electrons mostly with microwaves) absorbances. NMR techniques (the same method as an MRI) measure the resonance frequency of the nuclei that form the molecule under a strong magnetic field and provide information about the location and distribution of the atoms.

Teachers' Guide 2

In every water molecule, the two hydrogen atoms bond to oxygen in such a way as to make a "mouse head"-shaped molecule, as represented in **Figure 2.2**. This shape allows a positively charged hydrogen atom from one water molecule to move very close to the negatively charged oxygen atom from another water molecule, as shown in **Figure 2.2**. When this happens, the two water molecules are held together by what is referred to as a hydrogen bond. It is important to understand that a hydrogen bond only occurs between water molecules. The bond that holds hydrogen and oxygen together within a water molecule is a different type, called a covalent bond.

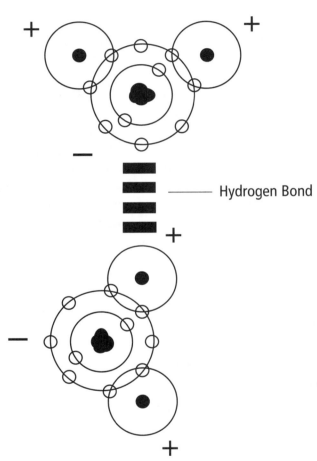

Hydrogen Bond

Figure 2.2
This illustration shows hydrogen bonding between two water molecules.

Although hydrogen bonds are relatively weak, they are strong enough to influence many of the physical properties of water. For example, they are partially responsible for water's high heat capacity (see Activity 4) and its high surface tension. This Activity demonstrates the polar nature of the bonds *within* a water molecule and the interaction *between* water molecules that results from hydrogen bonding.

Preconceptions

Ask students to tell you what they think they know about water molecules. Because the shape of the molecule is as important as its constituents, you can ask them to draw or describe what they imagine the molecule looks like. Here is a possible preconception students might have:

- Matter is continuous; it is not made of particles that act as building blocks.

What Students Need to Understand

- The model students construct is a two-dimensional (flat) representation of a three-dimensional molecule.
- The polar characteristics of water cause attractive forces between water molecules, which help account for many of the unique properties of water.

Time Management

The construction of the paper model and class discussion may be completed in about 50 minutes.

Preparation and Procedure

Copy one molecule pattern sheet per student; construction paper or card stock is recommended. The hydrogen and oxygen atoms may be copied on different colors of paper, or students may color their molecules.

Be sure to include a discussion relating to the information in What Is Happening? before asking students to answer questions 3 through 6, but *after* they have arranged the nuclei and electrons on their own.

Be prepared to point out to students that the "sticking together" that they model in response to question 6 results from hydrogen bonds, and that these bonds account for water's high heat capacity, high surface tension, and other properties.

Extended Learning

- Have students build physical models of water molecules. Since these models (of clay or from kits) are three-dimensional, they will give students a more realistic sense of the molecules.
- Explore the arrangement of water molecules in the solid, liquid, and gaseous states. Use several students' models—if students have made three-dimensional models, use them—to show how water molecules are arranged in an ice crystal. Simulate the motion of water molecules in each of its three phases. (See **Figure 2.3**.) You may want students to save their molecules for discussion in Activity 3.

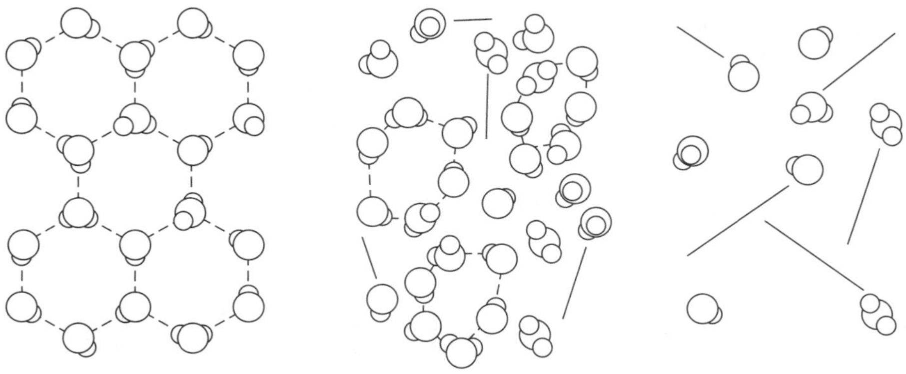

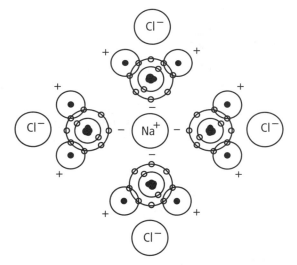

Solid
(many hydrogen bonds)

Liquid
(some hydrogen bonds)

Gas
(no hydrogen bonds)

Figure 2.3
This illustration shows the bonding and relationship of water molecules in the solid, liquid, and gaseous phases.

Students may wish to make paper models of substances commonly found in ocean water, such as sodium chloride (NaCl). Students can then demonstrate how water molecules are arranged around both positive and negative ions when such a substance dissolves. (See **Figure 2.4**.) (Also see Reading 1: Water: The Sum of Its Parts for a more detailed explanation of molecular structure. Activity 5 might also be helpful.)

Figure 2.4
This illustration shows the relationship of dissolved sodium (Na$^+$) and chlorine (Cl$^-$) ions in relation to water molecules.

Interdisciplinary Study

People did not always explain water's "stickiness," its shape, or how the atoms in a water molecule bond in the same way that we do today. Investigate the work of Robert Hooke, Antonie van Leeuwenhoek, and Robert Boyle, and see how their observations and explanations of water and its structure compare with our current explanations. Some of their ideas might seem silly today—like Hooke's belief that bonds were created between molecules by hooks from one molecule intertwining with the loops of another—but encourage students to think about why such a theory seemed reasonable to those scientists.

Differentiated Learning

- Some students will benefit from seeing and manipulating physical models of water molecules or animations, such as those by the American Chemical Society. Search for "ACS Middle School Chemistry multimedia Chapter 2" at *www.acs.org*. Go to the animations of Lesson 2.
- Another source, which animates hydrogen bonding, is the National Science Foundation Special Report: The Chemistry of Water. Search for "NSF News Chemistry Water" at *www.nsf.gov/news/special_reports/water/index_low.jsp*. In the Overview, go to "Attractive Forces."

Answers to Student Questions

1. The elements found in water are hydrogen and oxygen.

2. The ratio of hydrogen atoms to oxygen atoms in a water molecule is 2:1.

3. (Be sure you have discussed covalent bonding before you assign this question.) The atoms of hydrogen and oxygen are covalently bonded. Each of the two hydrogen atoms *share* one electron with the oxygen atom.

4. When the ends of a molecule are positively and negatively charged, the molecule is polar. Water is a polar molecule because the electron of each hydrogen atom is closer to the oxygen atom when hydrogen and oxygen are bonded together, making the oxygen end of the molecule negative and the hydrogen end positive.

5. They will tend to attract each other.

6. The positive and negative ends of the water molecules are attracted to one another and cause the molecules to stick together. (Note: Students should be encouraged to think of a variety of ways in which two water molecules could orient themselves. Have them work with other students on this question. The correct orientation is shown in **Figure 2.2**.)

7. The molecules move around, and some will stick in this configuration at least temporarily because the positive pole on one will be attracted to the negative pole on another.

Assessment

- You can ask students to prepare for friends in other classes an explanation of how the shape of a water molecule affects the way it bonds. They could write their explanation or draw cartoons showing it. You could use this— or the responses to questions—as a formal summative assessment.
- You could ask students to write a movie review of "Molecule Profile: H_2O–Water" (4:46) from NBC LEARN's video archives. (If you plan to do Activity 5, stop the video at 3:10.) Search for NBC LEARN K–12 at *http://archives.nbclearn.com/portal/site/k-12*.

Connections

Water is not the only molecule with covalent bonding and a shape that makes it polar. Students can learn about ammonia/methane, another polar molecule that is common in the solar system and that was abundant in Earth's early atmosphere. Saturn's moon, Titan, has an atmosphere with high methane content. Scientists have suggested that a form of life may exist on the surface, using liquid methane as a medium instead of water. See *http://solarsystem.nasa.gov/scitech/display.cfm?ST_ID=1389*.

Resources

http://ga.water.usgs.gov.edu

http://solarsystem.nasa.gov/scitech/display.cfm?ST_ID=1389

www.acs.org

www.nsf.gov/news/special_reports/water/index_low.jsp

http://archives.nbclearn.com/portal/site/k-12

Activity 3 Planner

Activity 3 Summary

Students drop pieces of solid gelatin, shortening, and water into their liquid phases to learn that water is an odd substance: Its solid phase is less dense than its liquid phase.

Activity	Subject and Content	Objective	Materials
Over and Under— Why Water's Weird	Density of solid versus liquid states of matter	Observe the behavior of different substances, including water, in their solid and liquid states.	Each group will need: several cups of ice cubes, several cups of unflavored gelatin cubes—clear and colored, vegetable shortening, teaspoon, spoon or tongs, hot plate with wire gauze screen, three 250 ml beakers , indirectly vented chemical splash goggles, gloves, aprons

Time	Vocabulary	Key Concept	Margin Features
30–50 minutes	Density	I: Properties of water	Safety Alert!, Fast Fact, What Can I Do?, Connections, Resources

Scientific Inquiry	Unifying Concepts and Processes
Observing	Structure and properties of matter

3

Over and Under— Why Water's Weird

Background

As you learned in Activity 1 and Activity 2, understanding water's molecular structure helps explain some of its characteristics. Now let's look at the **density** of water. Density, as you might recall from other science courses or investigations, is the mass of an object divided by its volume. Imagine dropping a small piece of wood in a glass of water. It floats because wood is almost always *less* dense than water. Now imagine dropping a piece of metal the same size as the piece of wood—a ball bearing, for example—into the same glass of water. The metal sinks. That's because it is *more* dense than water. (See **Figure 3.1**.) This example examines three different substances—wood, metal, and water—but even the same substance can have different densities depending on its temperature.

Vocabulary

Density: The heaviness of objects of similar size. Formally, density is the mass of an object divided by its volume.

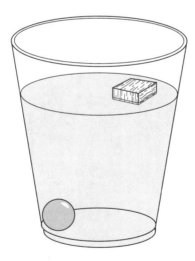

Figure 3.1
The wood is less dense than water, while the metal ball bearing is denser than water; one floats and the other sinks when placed in a cup of water.

Topic: water
Go to: *www.scilinks.org*
Code: PESO 003

Objective
Observe the behavior of different substances, including water, in their solid and liquid states.

Activity 3

Figure 3.2
Materials needed for the Activity

In general, materials become denser when they are cooled and less dense when they are heated. Think of hot air rising and cool air sinking from home heaters along the floor and air conditioners high on a wall. We also know that, generally, when we heat a solid to a high enough temperature, it forms a liquid. (If heated to a higher temperature, this liquid will change into the third state of matter—gas.) It follows then that as most materials are cooled, they become denser because their mass is constant but they contract. When warmed, these materials become less dense because their mass is constant but they expand.

In this experiment, we will look at how heating affects the density of different substances.

Procedure

1. Have students observe wood floating on water and a penny sinking in it. As a lead-up to the gelatin/shortening experiments, this should reinforce the idea that less dense objects float and more dense objects sink.

2. Place one of the beakers on the gauze screen on the hot plate. (See **Figure 3.2**.) Using the spoon or tongs, put several cubes of the clear gelatin in the uncovered beaker. Heat over medium heat until they just melt. (Use medium heat so that the cubes—in this step and the rest—melt slowly enough for you to observe.)

3. Using a spoon or tongs, carefully place a small cube of colored gelatin into the melted gelatin. Note whether it sinks or floats.

4. Repeat this procedure with the vegetable shortening. Slowly heat several teaspoons of the shortening until just melted. Then carefully place an additional teaspoon of the solid shortening in the newly formed liquid. Note whether it sinks or floats.

5. Repeat this procedure with ice. Slowly heat the ice cubes until just melted. Then carefully place an ice cube in the newly formed water. Note whether it sinks or floats.

Questions and Conclusions

1. Did the solid gelatin float or sink when you added it to the liquid gelatin? Why?

2. Did the solid vegetable shortening float or sink when you added it to the liquid shortening? Why?

3. Did the solid water (ice) float or sink when you added it to the water? Why?

4. Most forms of matter expand when heated and contract when cooled. How is water different?

5. How would life in ponds, lakes, and oceans be affected if frozen water behaved like other frozen substances?

6. Why do you think your teacher had you melt ice and put the ice cube in it, instead of just putting a cube in tap water?

Fast Fact

At 0°C, when freshwater freezes, its volume increases by about 9%.

Floating ice caused the deaths of 1,589 people on April 15, 1912 when the *RMS Titanic* struck an iceberg (a floating chunk of glacier) and sank. That happened a century ago, but on November 24, 2007, the *MV Explorer* struck an iceberg near Antarctica and sank. This time, all passengers aboard were rescued.

What Can I Do?

Every winter, someone's water pipes freeze. It is an inconvenience (no water) and a potential disaster (a flood inside the building when the pipes thaw). The problem is a consequence of water's expansion as it freezes. Learn how you can prevent pipes from freezing, what you should do if they freeze, and what you should not do if they freeze (i.e., thaw them with a blowtorch).

Over and Under— Why Water's Weird

What Is Happening?

The solid form of a substance is usually denser than the liquid form; therefore, the solid form of a substance generally sinks in the liquid form. Water, however, is peculiar. Like other substances, water expands when heated and contracts when cooled. Yet, when water is cooled below 4°C, it begins to expand rather than contract. At 0°C, freshwater freezes into a solid that is less dense than water that is slightly warmer than 0°C. (See **Figure 3.3**.) The result is that ice takes up more space than an equal mass of liquid water. Because ice is less dense than water, it floats.

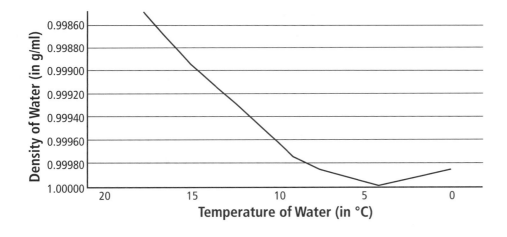

Objective

Observe the behavior of different substances, including water, in their solid and liquid states.

Key Concept

I: Properties of water

Materials

Each group will need

- several cups of ice
- several cups of unflavored gelatin cubes—clear and colored
- vegetable shortening
- teaspoon
- spoon or tongs
- hot plate with wire gauze screen
- three 250 ml beakers
- indirectly vented chemical splash goggles
- gloves
- aprons

Time

30–50 minutes

Figure 3.3 Density variation of water with temperature

How Do We Know This?

We explain the fact that ice floats on water by its open structure as a solid. How do we know what its structure is?

As in the case of the water molecule in Activity 2, we can use spectroscopic techniques to get information about the structure of ice, but the most common technique uses the diffraction of different radiations (x-rays, neutrons). For instance, x-ray diffraction consists of the apparent bending of waves around the electron clouds of the atoms forming the ice crystal.

Preconceptions

Ask students if they have ever dropped a stick in a pond or stream, bay, or ocean. What happened? What about dropping a rock in the water? Then, ask them to explain why the wood floated and the rock sank. From this, you can learn what students do and don't understand about density. Or, ask students to write in a journal or describe in a Think-Pair-Share what they understand about density. The following are some preconceptions that students might have:

- The mass of a substance changes when it changes state.
- The temperature of ice does not change.
- Many solid substances such as volcanic rocks float on their liquid counterparts.

What Students Need to Understand

- While water is a common substance on Earth, its properties are unique.
- The solid form of most substances is denser than the liquid form. Therefore, the solid form sinks when placed in the liquid form of the substance.
- Water is unique in that it is denser as a liquid than as a solid. Water is densest at 4°C, while still in its liquid state.
- The fact that solid water floats in liquid water has significant implications for life on Earth.

Time Management

Students can complete this Activity in between 30 and 50 minutes. It also works well as a station activity with other similar activities that deal with water's unique properties.

Preparation and Procedure

Make the ice cubes and gelatin cubes similar in size and prepare them well in advance. Prepare two to four packets of unflavored gelatin, following the directions on the box. (Note: Use unflavored gelatin in this activity to emphasize that no one should eat laboratory materials.) Once the gelatin powder dissolves, divide the solution in half. To one half of the gelatin, add enough food coloring to make it dark. Leave the other half uncolored. Chill both batches of gelatin overnight, then cut into cubes. (Note: You can substitute paraffin wax for the gelatin cubes. However, use caution with hot, liquid wax.)

When selecting a vegetable shortening for this Activity, avoid using shortenings that have had air whipped into them during their preparation. Be sure to test the shortening you have chosen prior to the start of the class to be sure it produces the desired results.

SAFETY ALERT

1. Be careful when working with the hot plate—skin can be burned.

2. Wash hands with soap and water upon completing the lab.

3. Indirectly vented chemical splash goggles, aprons, and gloves are required.

4. Be careful to quickly wipe up any spilled water on the floor—slip and fall hazard.

Have students melt the gelatin, the shortening, and the ice slowly using a heat source set on medium. If the liquid is too hot, the ice and gelatin cubes will melt too quickly for observation.

Do this entire Activity yourself before having students do it. That will allow you to anticipate safety and logistical issues.

Extended Learning

- Ask students to explain how ships and boats made of dense steel are able to float on water.
- Have students think of other substances they have observed in which the solid form sinks in the liquid form. Ask them to think of other substances to test. (Note, however, that many substances will dissolve or be affected by water.)
- The fact that water expands when it freezes has important ramifications in geology and the rock cycle. Street paving is also affected by water's expansion when frozen. (What causes potholes?) Have students investigate these relationships through library or internet research, or through experimentation.
- Scientists surmise that both liquid water and ice exist on Saturn's moons, Enceladus and Titan, as well. A video from Teachers' Domain at *www. teachers domain.org*, "Ingredients for Life: Water," (4:44), introduces the evidence for water and the implications for life on these planetary bodies. Have students investigate the evidence for both states of water on Europa (Jupiter's moon) or other planetary bodies.
- Have students investigate the techniques we have used to detect the water.

Interdisciplinary Study

Have students investigate Earth's frozen water (e.g., icebergs, frozen lakes, etc.) and how it affects humans (e.g., shipping and transportation, recreation, etc.). They can, for instance, learn about proposals to solve water shortages in London and southern California by towing icebergs there.

Differentiated Learning

- Many students wrestle with the concept of density—or with the physical meaning of any ratio or proportion. To make density intuitive for students, have at least two samples of substances of the same volume but different masses (for example, an empty bottle, and the same-size bottle filled with water; an empty box, and the same-size box full of rocks). You might find other such materials with physical science supplies. Encourage students to *hold* the samples so they can *feel* the different masses. Alternatively, you can make piles of gelatin, shortening, and ice of the same mass for students to *see* that they have different volumes.

SAFETY ALERT

Review student safety procedures concerning working with a hot plate and heated materials. Students must wear indirectly vented chemical splash goggles, aprons, and gloves. For some classes, you may want to use this Activity as a demonstration.

Connections

Since the *Galileo* spacecraft revealed slabs of ice on Europa, a moon of Jupiter, planetary geologists have thought that a deep sea of liquid water might exist under the ice. For a short video about the discovery, search for "Ice Sheets on Jupiter's Moon" (3:34) on the NBC LEARN K–12 site at *http:// archives.nbclearn.com/portal/ site/k-12*.

- For students who would benefit from a short digital review of density, search for "Inquiry in Action Chemistry Review: Density" at *www.inquiryinaction.org/chemistryreview/density*.
- For students who would benefit by visualizing water freezing on a molecular scale, search for "ACS Middle School Chemistry multimedia Chapter 2" at *www.acs.org*. Lesson 2 has videos on water molecules and water as a liquid. Lesson 4 has two videos about the structure of ice.
- For an animation that compares the structure of ice to that of liquid water, search for "Edinformatics ice float" at *www.edinformatics.com/interactive_molecules/ice.htm*.

Answers to Student Questions

1. The solid gelatin sank in the liquid gelatin. Gelatin in the solid form is denser than in the liquid form. Therefore, the solid sinks.

2. The solid shortening sank in the liquid shortening. Shortening in the solid form is denser than in the liquid form. Therefore, the solid sinks.

3. The ice floated on the surface of the water. Solid water, ice, is less dense than liquid water and, therefore, it rises in the water and floats on the top.

4. Water expands when it is heated and also when it freezes. (Water is most dense at 4°C.)

5. Plants and animals that live in the bottom of a pond or lake would be frozen in the water during the cool winter months.

6. Scientists should follow the same procedure for all of the substances; it is good experimental practice that allows for valid comparisons.

Assessment

- You can assess prior student knowledge of the density of water by asking the questions included in Preconceptions.
- You can assess formatively, guide students, and troubleshoot by circulating as students do this Activity. You can ask them to write down any questions that arise as they work, and discuss those as a class at the end of the Activity. They might want to know, for instance, why water is different from the other substances. (See How Do We Know This?)

- For an authentic assessment, ask students to explain the demonstrations in "Ice Bomb," a five-minute video on Teachers' Domain at *www.teachers domain.org*. You can recreate the first demonstration or show the video. You should not try to recreate the second demonstration, but it *will* demonstrate why pipes burst.
- For formal summative assessment, you can grade answers to questions.

Resources

http://archives.nbclearn.com/ portal/site/k-12

www.teachersdomain.org

www.inquiryinaction.org/ chemistryreview/density

www.acs.org

www.edinformatics.com/ interactive_molecules/ice.htm

Activity 4 Planner

Activity 4 Summary

Students explore specific heat by heating equal masses of sand and water under a lamp, measuring the temperatures of both, and recording observations. They then let the same samples cool. Students then graph and analyze their results.

Activity	Subject and Content	Objective	Materials
How Water Holds Heat	Specific heat of sand and water	Compare the specific heat of sand and water.	Each group will need: two thermometers (nonmercury), ring stand, two utility clamps, two 240 ml (8 oz.) polystyrene foam cups, sand, water, string, lamp with reflector and 200-watt bulb, clock or watch with a second hand, balance, graduated cylinder, ruler or meter stick, GFI-protected circuit to plug the lamp into

Time	Vocabulary	Key Concept	Margin Features
50 minutes	Specific heat, Calorie	I: Properties of water	Fast Fact, Safety Alert!, What Can I Do?, Connections, Resources

Scientific Inquiry	Unifying Concepts and Processes
Measuring and graphing data	Measuring relative rate of change

How Water Holds Heat

Background

Imagine that you put a cold iron skillet and a pot of cold water on the stove at the same time. You turn the burners underneath each of them on high. After a few minutes, would you be more willing to touch the skillet or put your hand in the water? Most people would be more willing to put their hand in the water than touch the skillet, because they know that some things heat up more quickly than others. This difference between substances results from a difference in **specific heat**, defined as the amount of energy required to raise the temperature of 1 g of a substance by 1°C. In the case of liquid water, that energy is equal to 1 **calorie**.

A substance with a high specific heat must absorb much energy before its temperature increases, and it must lose a lot of energy before its temperature decreases. In contrast, when a substance has a low specific heat, it takes relatively little energy to change the temperature of the substance.

Earth's surface is mostly covered by water. Areas not covered by water—the continents—are surrounded by water. If land and water had the same specific heat, we would expect the land and surrounding water to heat up and cool down at the same rate. From experience, we know this is not the case. In this Activity, we will investigate the heat capacities of sand and water as a simple model of land and a body of water.

Vocabulary

Specific heat: The amount of heat energy (measured in "calories") required to raise the temperature of 1 g of a substance by 1°C.

Calorie: The amount of energy necessary to raise the temperature of 1 g of water by 1°C.

Fast Fact

Temperatures on the surface of the Moon swing wildly from one extreme to the other, from an average of −184°C (−300°F) in the shade to an average of 101°C (214°F) in the Sun.

Topic: water properties
Go to: *www.scilinks.org*
Code: PESO 002

Objective

Compare the specific heat of sand and water.

Activity 4

Materials

Each group will need

- two thermometers (nonmercury)
- ring stand
- two utility clamps
- two 240 ml (8 oz.) polystyrene foam cups
- sand
- water
- string
- lamp with reflector and 200-watt bulb
- clock or watch with a second hand
- balance
- graduated cylinder
- ruler or meter stick
- GFI-protected circuit to plug the lamp into

Time

50 minutes

SAFETY ALERT !

The lamp will become hot. Take care when using electricity around water. Contact the teacher immediately if you get burned.

Procedure

1. Using a balance, measure 200 g of sand and place it in one of the polystyrene foam cups.

2. Measure 200 ml of water in the graduated cylinder and place in the other polystyrene foam cup. (Note: You can measure the volume instead of the weight because 1 ml of water has a mass of 1 g. Thus, we are measuring 200 g of water as well.)

3. Attach the two utility clamps to the ring stand. (See **Figure 4.1**.)

4. Suspend the thermometers from the utility clamps with the string. Adjust the height of the clamp so that the bulb of each thermometer is just covered by the sand or water. The bulb of each thermometer should be in the center of the container. (See **Figure 4.1**.)

5. Position the lamp 30 cm above the two containers. Be careful to center the lamp so that each container receives an equal amount of light. The setup should look similar to **Figure 4.1**.

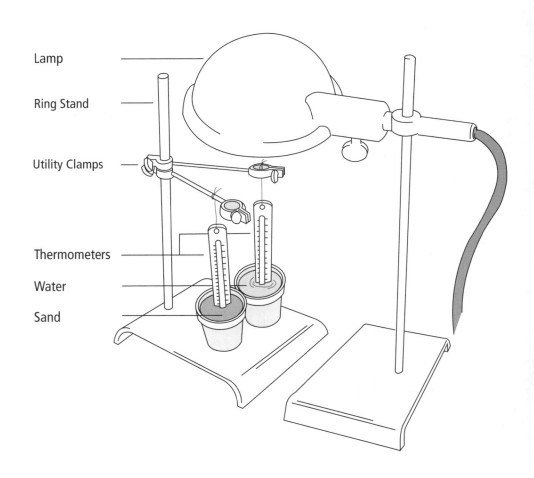

Lamp

Ring Stand

Utility Clamps

Thermometers

Water

Sand

Figure 4.1
Laboratory setup to measure the specific heat of water and sand

6. In the data table in **BLM 4.1**, record the starting temperatures of both the sand and water. (Note: The temperatures of the sand and water do not necessarily have to be the same.) Also, record comments and observations at each reading—such as "water temp. constant for last four min." or "water temp. decreasing 1°C per min."

7. Turn on the lamp and record in the table the temperature of the sand and water *every* minute for 15 minutes.

8. After 15 minutes, turn the lamp off. Continue to record the temperature of both the sand and water every minute for another 15 minutes.

9. Using the data from the table, plot the changes in the temperature of the sand and water over the 30-minute period. Use the graph provided in **BLM 4.2**.

Questions and Conclusions

1. What was the temperature change of the sand and the water after 15 minutes under the lamp?

2. In which substance did you observe the faster increase in temperature?

3. How many degrees did each substance cool in the 15 minutes after you turned the lamp off?

4. Which substance retained its heat energy and cooled more slowly?

5. If the sand and the water received the same amount of heat energy, why were there differences in the heating and cooling rates?

6. Describe how differences in the rates of heating and cooling of water and sand could cause the climate along the seacoast to differ from that further inland.

SAFETY ALERT

1. Wash hands with soap and water upon completing the lab.

2. Be careful when working with a hot lamp—skin can be burned. Notify your teacher immediately if someone is burned.

3. Be careful to quickly wipe up any spilled water on the floor—slip and fall hazard.

What Can I Do?

You can experience the effect of water's high specific heat by swimming in the fall. Note how the water feels compared to the air. You could measure the temperatures of both the air and water as weather changes or fall arrives. You can also explore the effect using data from several high-tech sources:

- NASA satellites and a Geographic Information System (GIS) such as Google Earth display sea surface temperatures. (You can download Google Earth for free.) To find the data, search for "NEO NASA Earth Observations" at *http://neo.sci.gsfc.nasa.gov/Search.html*. Select the Ocean tab near the map.
- National Data Buoy Center at *www.ndbc.noaa.gov*
- National Oceanographic Data Center at *www.nodc.noaa.gov/dsdt/cwtg*

Data Table: Temperature Comparison

Time (min.)		Sand Temperature	Water Temperature	Comments
Lamp On	1			
	2			
	3			
	4			
	5			
	6			
	7			
	8			
	9			
	10			
	11			
	12			
	13			
	14			
	15			
Lamp Off	16			
	17			
	18			
	19			
	20			
	21			
	22			
	23			
	24			
	25			
	26			
	27			
	28			
	29			
	30			

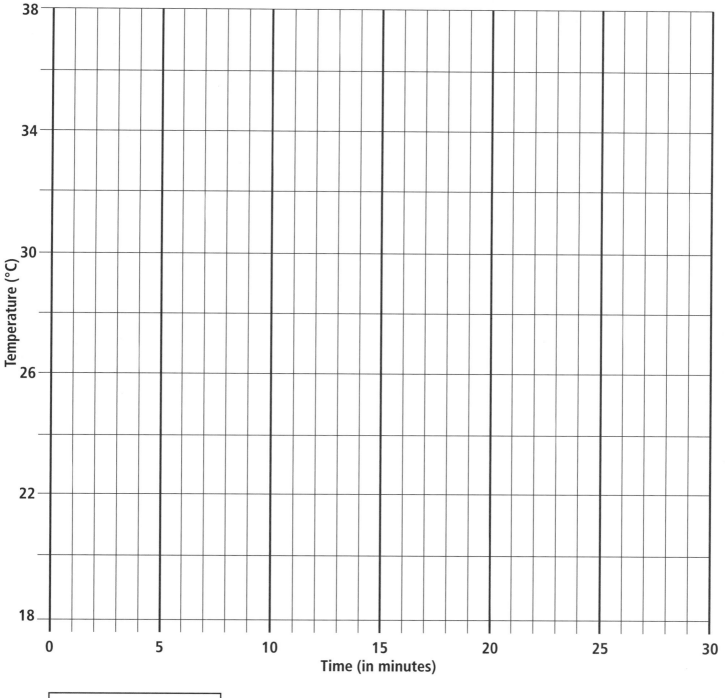

Change in Temperature of Sand and Water over Time

Key
- - - - - - Sand
————— Water

How Water Holds Heat

What Is Happening?

The specific heat of water is fundamental to the study of oceanography and to developing an understanding of the important role of water in shaping Earth. Specific heat is the amount of heat energy (measured in calories) needed to raise the temperature of 1 g of a substance 1°C. If you take a given mass of water and an equal mass of another substance at the same temperature, and heat them equally over time, you find the other substance, with two exceptions, becomes hotter than the water. Water heats up slowly because it has a high specific heat. In fact, apart from the two exceptions—hydrogen gas (3.4 cal/g) and liquid ammonia (1.23 cal/g)—water (1.0 cal/g) has the highest specific heat of all substances. (Sand is not a specific chemical compound but rather a mixture, so its specific heat varies. As a point of reference, the specific heat of sandstone is approximately 0.26 cal/g. This provides a rough comparison between sand and water.)

The specific heat of a substance depends on the size and weight of its atoms. The larger the atoms, the less heat required to raise the temperature. Specific heat is inversely proportional to a substance's atomic weight.

Due to the high specific heat of water, lakes and oceans change temperature slowly when heated by the Sun or cooled by the atmosphere. Therefore, large bodies of water act as buffers to adjacent landmasses and greatly affect climate. Coastal land areas have climates with smaller temperature ranges than those that occur in inland areas.

This Activity explores the concepts of specific heat by comparing the heating and cooling rates of sand and water.

Objective
Compare the specific heat of sand and water.

Key Concept
I: Properties of water

Materials
Each group will need

- two thermometers (nonmercury)
- ring stand
- two utility clamps
- two 240 ml (8 oz.) polystyrene foam cups
- sand
- water
- string
- lamp with reflector and 200-watt bulb
- clock or watch with a second hand
- balance
- graduated cylinder
- ruler or meter stick
- GFI-protected circuit to plug the lamp into

Time
50 minutes

How Do We Know This?

How do scientists estimate sea surface temperature from satellites?

Scientists estimate sea surface temperature from satellites by measuring the radiation emitted by the surface of the ocean. The same way that satellites can observe land in the visible part of the electromagnetic spectrum (check, for instance, Google Earth), some other satellites observe Earth at other frequencies, and some of them provide temperature information for the ocean. The most common satellites measure thermal infrared radiation, which is directly related to the temperature of any body. Some more recent satellites use microwaves to infer the temperature over the ocean.

Preconceptions

Ask students to describe what they recall about heat and temperature. Probe to see if they have learned about specific heat before and, if so, what they understand about it. You can ask them, for instance, "What do you think would happen to their temperatures if 100 g of gold and water were heated in two pans in a 250°F oven for two minutes?" Then discuss students' responses to discover what they really know about specific heat. The following are some examples of possible preconceptions:

- All materials, including water, change temperature at the same rate if heated (or cooled) in the same environment.
- Heat and temperature are the same thing.
- A phenomenon called cold exists that can be added to a substance to lower the substance's temperature, just as heat can be added to raise its temperature.

What Students Need to Understand

- Different materials absorb and release heat at different rates.
- The specific heat of water is high. Water's specific heat is responsible for the ocean's moderating effect on the climate of coastal regions.

Time Management

This Activity can be completed in 50 minutes. Allow 10 to 15 minutes to distribute and set up materials. The experiment itself requires 30 continuous minutes. Remember to allow time for students to disassemble materials. Make sure each class starts with freshwater and sand because the materials must cool to room temperature before students heat them again. You might need to extend this experiment into the next class to discuss the Activity thoroughly.

Preparation and Procedure

The sand should be thoroughly dry. Any amount of moisture may interfere with experimental results. Also, it is not critical that the starting temperatures of the sand and water are exactly the same, but it does make interpreting results much easier. You can equalize temperatures in the starting materials by storing the sand and water at room temperature overnight.

Since specific heat is defined as the amount of energy required to raise 1 g of a substance 1°C, it is important to use equal *masses* of sand and water, rather than simply filling the cups to the same volume. You may want to make students specifically aware of this point, especially if you spend time during your laboratory activities discussing the control of experimental variables and sources of experimental errors.

SAFETY ALERT

1. Wash hands with soap and water upon completing the lab.

2. Be careful when working with a hot lamp—skin can be burned.

3. Be careful to quickly wipe up any spilled water on the floor—slip and fall hazard.

4. Review student safety procedures concerning use of water and electricity, and for handling a hot lamp. Plug the lamp into a GFI-protected circuit only.

Extended Learning

- To reinforce the Sun's role in Earth's systems, consider doing this Activity outside, too. Set two buckets containing equal masses of sand and water outdoors in an open area with full exposure to sunlight. Record the starting temperature of both the sand and the water, and continue to take readings throughout the school day. Compare the results obtained using the lamp to the results obtained outdoors.

- Have students do this Activity with different kinds of sand—plus the water. ("Sand" strictly means a size range of mineral grains, not a specific composition.) They might be able to compare the temperature changes of quartz or carbonate sands to beach sand. As well, beach sand sometimes contains layers of dark minerals; students could use those layers as a separate sample. Quartz sand is usually for sale at building supply stores.

- Due to water's high specific heat, oceans hold vast quantities of heat and change their temperature slowly. Students can see how sea surface temperatures change seasonally using Geographic Information Systems (GIS). Both Google Earth and ArcGIS Explorer (software for download at no cost) can display NASA sea surface temperature data. Search for "NEO NASA Earth Observations" at *http://neo.sci.gsfc.nasa.gov/Search.html*. Select the Ocean or Energy tabs near the map to see data. For instance, the link to "Sea Surface Temperature 2002+ (MODIS)" allows students to get data month by month so they can compare January temperatures to July temperatures. They can display or download and save the data as KML files for Google Earth or ArcGIS Explorer.

- The concept presented in this Activity provides a good opportunity to relate oceanography and meteorology. Ask students to compare the climates along the coast to those inland, nearby, and at about the same elevation. They will probably want to compare the spread between the average daily high and daily low temperatures for two places at different times of year. For example, the difference between average maximum and minimum temperatures at Hatteras on the Outer Banks of North Carolina on July 1 is 13°F. At Smithfield, North Carolina, 85 miles inland, it is 23°F. For January 1 at Hatteras, the difference is 15°F, while at Smithfield, it is 22°F. Students can find the data by searching online for the Southeast Regional Climate Center or "SERCC NOAA NOWData" at *www.sercc.com/nowdata.html*. Despite the name, this source has data for all 50 states. Students will need to choose the region, the Daily/monthly normals, the specific locations, and the variables (e.g,. Max Temperature and Min Temperatures). Alternatively, they can select the Daily almanac as the product, the specific locations, and dates (e.g., January and July 1) for comparison. When the almanac pops up, they will probably want to examine the Normal Maximum Temperature and Normal Minimum Temperature. See Connections for further suggestions.

Connections

The difference between the specific heats of land and water influences coastal weather and climate and leads to several ways to investigate the ocean–atmosphere–geosphere connections:

- Ask students to experiment during their visits to lakes or oceans and try to relate their experiences with the ideas in this Activity (i.e., walking across hot sand, wading in cold ocean water, etc.).
- Have students investigate what "lake effect" is. Using newspaper weather maps or the local weather bureau for information, find out why there is more snow in the winter in certain areas around such large bodies of water as the Great Lakes.
- Have students investigate "sea breeze." What time of the day are sea breezes stronger? Why?

Interdisciplinary Study

- Have students investigate why more than half of the world's population lives along the coastline.
- Because of the climate of the Great American Desert, far from oceans, what hardships did the pioneers endure on wagon trains of the 1800s? Have students research conditions on the Oregon or California trails in the arid West in the 19th century.
- Have students compare the effects of different climates on the livelihoods of Native American tribes in interior regions with those of coastal tribes. How were these tribes' methods of finding sustenance and their customs influenced by the climate?

Differentiated Learning

You can ask students in advanced mathematics to graph their results of the Activity on a graphing calculator instead of on **BLM 4.2**. They can also determine the equation of the line that best represents their data. This is an activity in which concrete thinkers benefit from exploring an abstract concept with a hands-on approach.

Answers to Student Questions

1. Answers will vary. The temperature change of the water should be less than the temperature change of the sand.

2. The sand increased in temperature faster than the water.

3. Answers will vary. The temperature of the sand decreases faster than the temperature of the water.

4. The water cooled more slowly; its temperature decreased gradually.

5. Water and sand are made up of different molecules and have different specific heats. The specific heat of water is higher than the specific heat of sand.

6. Because water retains heat so well, the temperatures along the coast do not fluctuate as much as those further inland. The ocean moderates coastal climates.

Assessment

- For formative assessment, circulate as students do this Activity and ask them, "What are you seeing happening? Is it what you expected? Why or why not? Would you like to do this experiment differently? In what way?

Resources

http://neo.sci.gsfc.nasa.gov/Search.html

www.ndbc.noaa.gov

www.nodc.noaa.gov/dsdt/cwtg

www.sercc.com/nowdata.html

- For summative assessment, if you asked students this question in Preconceptions, ask them, "*Now*, what do you think would happen to the temperatures if 100 g of gold and water were heated in two pans in a 250°F oven for two minutes—and why do you think that?"

- You can ask students to write a paragraph that explains what they did in this Activity and what their results were. You can also ask them to then extend their lab results to the coast: If the Sun shines at the beach (or coast or shore), how does it affect the land compared to the water?

- You can grade students' answers to the questions.

Activity 5 Planner

Activity 5 Summary

Students observe how well salt, sugar, baking soda, and Epsom salts dissolve in water, rubbing alcohol, and mineral oil. They explore how the solubility of substances depends on the atomic arrangement or polarity of the solute and solvent.

Activity	Subject and Content	Objective	Materials
Water— The Universal Solvent	Solubility	Explore the solubility of various substances in water as compared with other liquids.	Each group will need: water, mineral oil (or baby oil), isopropyl alcohol (70%), Epsom salts (Mg_2SO_4), baking soda ($NaHCO_3$), table salt (NaCl), granular sugar, red wax marking pencil, five test tubes with rubber stoppers, small spoon (1/8 teaspoon will work), test tube rack, graduated cylinder, black construction paper

Time	Vocabulary	Key Concept	Margin Features
50 minutes	Solvent, Solubility	I: Properties of water	Safety Alert!, Fast Fact, What Can I Do?, Connections, Resources

Scientific Inquiry	Unifying Concepts and Processes
Observing and organizing data in organized way	Structure and properties of matter

Water— The Universal Solvent

Background

Water is often called the universal **solvent** because so many substances will dissolve in it. Why do so many substances dissolve readily in water?

The ability of a substance to dissolve in a liquid—its **solubility**—is dependent on the molecular arrangements of the liquid and the substance. Water is a polar molecule with positive and negative ends. (You might have learned this in Activity 2.) For this reason, other substances with positive and negative surface charges are attracted to the positive (+) and negative (−) ends of water and are therefore dissolved and kept "in solution."

Salts—one example is NaCl (sodium chloride), also known as table salt—readily dissolve in water because they contain positive and negative ions that are attracted to the water molecules. (See **Figure 5.1**.) Ocean water contains many types of dissolved salts, most of which originally came from rocks that had been altered or which decomposed. Some salts are picked up by rainwater and melting snow, are carried by the rivers, and eventually end up in the ocean. Others originate within the ocean when substances such as seashells dissolve in the ocean water, and when underwater hot springs add or remove chemicals from seawater. Still others make their way to the ocean through the atmosphere. However, the ocean does not grow continually saltier. Salts eventually crystallize and settle out of ocean water.

> ### Vocabulary
>
> **Solvent:** A substance (such as water) that can dissolve other materials.
>
> **Solubility:** The amount of a substance that will dissolve in a given amount of liquid.

Objective

Explore the solubility of various substances in water as compared with other liquids.

Topic: water properties
Go to: www.scilinks.org
Code: PESO 002

Activity 5

Figure 5.1
Salt (NaCl) dissolves readily in water.

Materials

Each group will need

- water
- mineral oil (or baby oil)
- isopropyl alcohol (70%)
- Epsom salts (Mg$_2$SO$_4$)
- baking soda (NaHCO$_3$)
- table salt (NaCl)
- granular sugar
- red wax marking pencil
- five test tubes with rubber stoppers
- small spoon (1/8 teaspoon will work)
- test tube rack
- graduated cylinder
- black construction paper
- indirectly vented chemical splash goggles
- gloves
- aprons
- MSDSs for all hazardous materials

Time

50 minutes

SAFETY ALERT

1. Indirectly vented chemical splash goggles, gloves and aprons are required.

2. Wash hands with soap and water upon completing the lab.

3. Pertinent safety procedures and required precautions outlined in MSDSs for hazardous materials will be reviewed with you by your teacher.

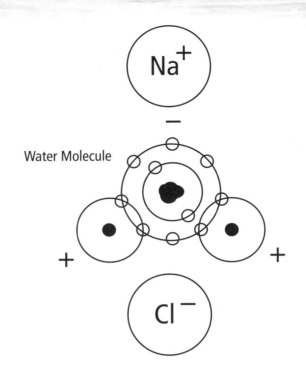

Water Molecule

Procedure

1. Put on indirectly vented chemical splash goggles, gloves and aprons.

2. Number the five test tubes with a pencil or red wax marking pencil. Pour 10 ml of water into the first four test tubes.

3. Measure 1/8 teaspoon of table salt and place it in test tube #5. This test tube will serve as a guide called control, showing how much solid material you started with.

4. Measure 1/8 teaspoon of table salt, and place the sample into the first test tube. Do NOT stir or shake the test tube. Watch the test tube for 1 minute. Record your observations in the table in **BLM 5.1**, estimating the amount of substance that has been dissolved by comparing it with the control test tube (#5) and any other comments about the mixture (clear, cloudy, etc.).

5. Cover the opening of the test tube with the rubber stopper; shake up and down vigorously 10 times. Look at the solution and record your observations. Place a piece of black construction paper behind the test tube to help you determine how much of the material has dissolved. Compare it with test tube #5.

6. Repeat step 4 until all the salt has been dissolved or until the particles will clearly not dissolve (50 shakes maximum). Record your observations after each set of 10 shakes.

7. Use the remaining three test tubes to repeat steps 3 through 5 for each of the following substances: granular sugar, baking soda, and Epsom salts. Record your observations in the table provided (**BLM 5.1**).

8. Repeat the procedure using isopropyl alcohol instead of water in the four test tubes. Record your observations from each trial in the table in **BLM 5.2**.

9. Repeat the procedure using mineral oil instead of isopropyl alcohol in the four test tubes. Record your observations from each trial in the table in **BLM 5.3**.

Questions and Conclusions

1. Which substance, or solute, dissolved fastest in water?

2. Which liquid, or solvent, did not dissolve any substance?

3. Explain why salts readily dissolve in water.

4. Why does shaking increase the amount of substance dissolved?

5. What are some common materials that we dissolve in water?

6. Why is water called the universal solvent?

7. What happens to materials in water that are not dissolved?

8. What are some common materials that are difficult to dissolve in water? Why?

Fast Fact

Table salt (sodium chloride) is one of thousands of compounds called salts. All of them are the product of an acid reacting with a base. For example, sodium chloride is the result of reacting hydrochloric acid (HCl) with sodium hydroxide (NaOH), yielding sodium chloride (NaCl) and water (H_2O).
$HCl + NaOH \rightarrow NaCl + H_2O$

What Can I Do?

If you live at the coast with its salt spray, or where people melt snow and ice with salt, you can see that salts can harm plants. You could design and do experiments to determine the effects of different salts on plants. You can also learn which plants are salt tolerant and would thrive near the beach or along a roadway.

Table: Solubility in Water

Test Tube	Substance	After 1 min.	10 Shakes	20 Shakes
#1	NaCl			
#2	Sugar			
#3	Baking Soda			
#4	Epsom Salts			
Test Tube	**Substance**	**30 Shakes**	**40 Shakes**	**50 Shakes**
#1	NaCl			
#2	Sugar			
#3	Baking Soda			
#4	Epsom Salts			

Table: Solubility in Isopropyl Alcohol

Test Tube	Substance	After 1 min.	10 Shakes	20 Shakes
#1	NaCl			
#2	Sugar			
#3	Baking Soda			
#4	Epsom Salts			
Test Tube	Substance	30 Shakes	40 Shakes	50 Shakes
#1	NaCl			
#2	Sugar			
#3	Baking Soda			
#4	Epsom Salts			

Table: Solubility in Mineral Oil

Test Tube	Substance	After 1 min.	10 Shakes	20 Shakes
#1	NaCl			
#2	Sugar			
#3	Baking Soda			
#4	Epsom Salts			
Test Tube	Substance	30 Shakes	40 Shakes	50 Shakes
#1	NaCl			
#2	Sugar			
#3	Baking Soda			
#4	Epsom Salts			

Water— The Universal Solvent

What Is Happening?

As discussed in previous Activities, water is chemically unusual. The fact that it is a polar molecule accounts for all of the properties investigated thus far: surface tension, heat capacity, and the relative densities of its solid and liquid form. Polarity is also responsible for water's dissolving ability. Water has the ability to dissolve a variety of substances, and thus water in the ocean contains large amounts of dissolved materials. Every element that occurs on Earth is present in seawater at least in small concentrations.

The solubility of substances, especially those that are ionic, is usually much higher in water than in other solvents. The solvent properties of water arise from the polarity of its water molecules—they have positive and negative ends. (See Activity 2.) The solute ions are surrounded by the oppositely charged polar ends of the water molecule, causing them to stay apart and not react together to form salt crystals. (See **Figure 5.2.**)

This Activity allows students to investigate the solubility of several different substances in three different solvents. Of the three solvents, water has the highest degree of polarity, which accounts for the increased solubility of the substances in water. In addition to the polarity of the solvent molecules, there are other factors that can affect solubility. These include the temperature of the solvent and the degree of mixing of the solute in the solvent. Temperature is not part of this Activity; however, students are asked to investigate the differences in relative solubility due to mixing.

How Do We Know This?

How does soap work?

Soaps and most detergents consist of long, thin molecules. One end of the molecule has a polar structure (**hydrophilic**): it loves water but hates oily liquids. The other end of the molecule is the opposite (**hydrophobic**): it likes oil but hates water. The hydrophobic part of the molecule surrounds the oil while the hydrophilic part forms hydrogen bonds with the water molecules. When we do laundry, the hydrophobic ends of the detergent molecules stick to the surface of the oily dirt on clothes, leaving the hydrophilic ends of the detergent surrounded by water; therefore, the oily dirt can be washed away.

Objective

Explore the solubility of various substances in water as compared with other liquids.

Key Concept

I: Properties of water

Materials

Each group will need

- water
- mineral oil (or baby oil)
- isopropyl alcohol (70%)
- Epsom salts (Mg_2SO_4)
- baking soda ($NaHCO_3$)
- table salt (NaCl)
- granular sugar
- red wax marking pencil
- five test tubes with rubber stoppers
- test tube rack
- small spoon (1/8 teaspoon will work)
- graduated cylinder
- black construction paper
- indirectly vented chemical splash goggles
- gloves
- aprons
- MSDSs for all hazardous materials

Time

50 minutes

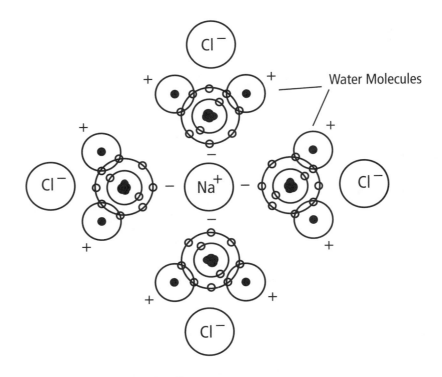

Figure 5.2
Common salt (NaCl) dissolved in water. Note the distribution of positive and negative charges.

Preconceptions

Ask students, "What have you observed when you have dissolved things?" You might also ask them to describe or draw what they imagine happens at the atomic scale when substances dissolve. The following are some examples of possible preconceptions.

- All liquids have the same ability to dissolve solids.
- All solids have the same chemical structure, even though their constituents differ.
- When a solid dissolves in water, it melts into the water.
- Dissolved substances eventually settle out.

What Students Need to Understand

- The solubility of a substance in a given solvent is dependent on the molecular arrangement of the solvent and substance involved.
- Water is a powerful solvent, dissolving many substances with which it comes in contact. This ability to dissolve many things comes from the fact that water is a polar molecule.
- Solubility changes with several factors, including mixing and temperature.
- River runoff is only one source of the salts found in the ocean. Other sources include underwater hot springs and submarine volcanoes, and dust from the atmosphere.

Time Management

This Activity can be completed in 50 minutes. Students will probably need the entire 50 minutes to do the laboratory part of the Activity and clean up, especially if you circulate and give them the opportunity to modify the procedure (see Assessment). Students will need additional time to discuss answers to questions in their groups, and to create a concept map on solubility if you assign it as an assessment.

Preparation and Procedure

Make sure the temperatures of the solvents are the same by storing all solvents for some time at room temperature. Prior to the Activity, be certain to review with students critical safety information from MSDSs for hazardous materials used in this lab (e.g., alcohol, etc.).

Extended Learning

- Have students expand their experimentation to test the solubility of other substances in water and to determine how changes in solubility result from changes in temperature or grain size.
- Have students research the atomic structure of each of the solvents and solutes. Ask them to look for relationships between structure and solubility. After completing this research, you may ask students to predict the solubility of other substances based on atomic structure.
- Explore the solubility of oils and fats in water. Do these compounds dissolve more easily in other substances, such as mineral oil or alcohol? Why? What happens to the solubility of oils in water when soap is added? What are the environmental implications of the modification in solubility (e.g., for oil spills)?
- Investigate the different ways that groundwater or ocean water can become polluted. Include mechanisms that involve dissolving pollutants in water, as well as pollutants that do not dissolve in water.

Interdisciplinary Study

- Have students investigate the chemistry of tears, sweat, urine, and sports drinks. They could compare the substances excreted to those replenished with commercial "hydration products." They could also explore the marketing and economics associated with those products.
- Students could also investigate the history of salt as a precious commodity that was used for preserving and seasoning food and, reputedly, as salary for Roman legionnaires.

SAFETY ALERT

1. Indirectly vented chemical splash goggles, gloves, and aprons are required.

2. Wash hands with soap and water upon completing the lab.

3. Review with students pertinent MSDS safety procedures and required precautions for hazardous materials.

Teachers' Guide 5

Connections

Geologists find many sedimentary rocks on Earth that formed from minerals that crystallized when saline solutions evaporated. Similar deposits on Mars provide more evidence that liquid water has been present there. Have students look into these "evaporite deposits" on Earth and Mars. They might already be familiar with some: gypsum (hydrated calcium sulfate) is the filler in sheet rock or wallboard. Hydrated sulfates exist on Mars. Search for "Evaporites Mars" at *www.newscientist.com/ article/dn11332-surging-groundwater-solves-martian-evaporite-mystery.html*.

• Have students bring in different brands of sea salts and compare them. Ask students to explore where they come from, how they are harvested, and how they are used in different kinds of cooking. This introduces a multicultural component to the Activity.

Differentiated Learning

This is a concrete Activity in which students make observations and draw conclusions.

• From their experimental results, students with a mathematical bent could calculate the maximum mass of salt that they could obtain by evaporating 1 m^3 of seawater. They would need to have access to a balance to determine the mass they added to reach saturation. (Average seawater is 3.5% salt by mass, but keep this secret until students have calculated a figure.)

• For students who need a visual image of the process of dissolving, you can show animations by the American Chemical Society. Search for "ACS Middle School Chemistry multimedia Chapter 5 Lesson 3" at *www.middleschoolchemistry.com/multimedia/chapter5/lesson3*. Lessons 3 through 9 pertain to dissolution.

Answers to Student Questions

1. Answers may vary depending on how coarse the substance's grains are and the temperature of the water, neither of which is controlled in this simple Activity. If students' results do vary within the class, this is an opportunity to guide a discussion about controlling variables and for the class to propose other experiments. Students will compare results with others, and will probably want to follow up if you encourage them to do so.

2. Baby oil or mineral oil did not dissolve any substance.

3. Salts readily dissolve in water because they contain positive and negative ions, which are surrounded by the molecules of water.

4. Without mixing, the substance is at high concentration at the bottom of the container, and the remaining material is less likely to dissolve completely. With mixing, the concentration of the substance is uniform throughout the solvent, and the concentration of the substance at the bottom of the container is decreased. The remaining material is more likely to dissolve.

5. Answers will vary but may include foods (drink mix, powdered milk, instant chocolate, etc.), medicines (Alka-Seltzer, headache powders, etc.), and others.

6. Water is called the universal solvent because so many substances dissolve readily in water.

7. Answers will vary but may include the following: Materials in water that are not dissolved sink to the bottom of the container or body of water, float on top of the water, or stay in suspension. (Note: This question provides a good opportunity to talk about water pollution. However, do not forget about the pollutants that do dissolve in water.)

8. Answers will vary, but may include the following: Some common materials that are difficult to dissolve in water are oils, fats, flour, and more solid materials such as metals (steel, iron, aluminum), sand, rocks, and others. Usually, these substances are not formed by ions, and therefore the polar structure of water is less effective as a solvent. Alteration of these substances can result in increased solubility: metals usually need to be oxidized to dissolve; soaps help dissolve oils.

Assessment

- While students are doing this Activity, ask them about their results. Have there been any surprises? What was different from what they expected? Would they like to modify the procedure? If so, how and why? Are there other substances they want to try?

- For formal assessment, you could ask students to create a concept map about solubility.

- You could also grade the answers to students' questions.

Resources

www.newscientist.com/
article/dn11332-surging-
groundwater-solves-martian-
evaporite-mystery.html

www.middleschoolchemistry.
com/multimedia/chapter5/
lesson3

Activity 6 Planner

Activity 6 Summary

Students make a hydrometer—an instrument to measure the relative density of liquid—from a plastic transfer pipette or straw and BBs. They then use the hydrometer to explore the effect of adding salt to the density of water.

Activity	Subject and Content	Objective	Materials
Won't You BB My Hydrometer?	Measuring density of freshwater and salt water	Understand the implications of water density by building and using a hydrometer to measure the densities of freshwater and saltwater samples.	For Part 1, each group will need: plastic transfer pipette, sharp scissors, fine-tip permanent marking pen, metric ruler, 20 BBs, 500 ml beaker, masking tape, modeling clay, food coloring, pickling salt, waste container, towels or rags for cleanup, teaspoon (5 ml) or tablespoon (15 ml) For Part 2, each group will need: large jar or beaker, pickling salt, soupspoon (large enough to hold an egg), 100 ml graduated cylinder, hard-boiled egg, ruler or straight-edge, 5 ml metric measuring spoon (a teaspoon), hydrometer (from Part 1)

Time	Vocabulary	Key Concept	Margin Features
90–100 minutes (45–50 minutes for each part)	Hydrometer, Density	I: Properties of water	Safety Alert!, What Can I Do?, Fast Fact, Connections, Resources

Unifying Concepts and Processes	Technology	Personal/Social Perspectives
Measuring and instrumentation	Building a hydrometer	Improvements in measuring devices allow science to advance.

Won't You BB My Hydrometer?

Background

If you have ever gone swimming in an ocean, or better yet, in Great Salt Lake, Utah, you may have noticed that it was easier to float in the ocean or in Great Salt Lake than in a pool or freshwater lake. Why is this? In Activity 5, we learned that many substances dissolve easily in water. Ocean water and the water in Great Salt Lake contain large amounts of dissolved salts.

In the first part of this Activity, you will construct a **hydrometer**—a device that allows you to compare the densities of different liquids. The word *hydrometer* means water (hydro) measure (meter). In this Activity, you will use your hydrometer to investigate how the addition of salts affects the **density** of water.

In the second part of this Activity, you will use your hydrometer and an egg to monitor the effects of salt in water.

Vocabulary

Hydrometer: An instrument that measures the relative densities of liquids.

Density: The heaviness of objects of similar size. Formally, density is the mass of an object divided by its volume.

Fast Fact

If you evaporated the water from 1,000 g of seawater, you would get 35 g of salts from just about anywhere on the open ocean. If you did the same at Great Salt Lake, you could get from 50 to 270 g of salts, depending on where you collected your sample.

Topic: water properties
Go to: *www.scilinks.org*
Code: PESO 002

Objective

Understand the implications of water density by building and using a hydrometer to measure the densities of freshwater and saltwater samples.

Activity 6

Figure 6.1
Cut approximately 2.5 cm off the tip of the pipette as shown.

Figure 6.2
Starting at 0 cm, mark ½-cm intervals on the masking tape as shown. Then wrap the tape around the pipette with the 0 cm mark at the tip.

Procedure

Part 1

1. Pick up a tray of materials from your teacher.

2. With a pair of scissors, enlarge the opening of the pipette by cutting approximately 2.5 cm off the tip. (See **Figure 6.1**.)

   Try to put a BB into the pipette. If the BB fits easily through the opening, proceed to step 3. If the BB sticks in the opening, cut off more of the pipette's tip. When the BB rolls into the pipette easily, proceed to step 3.

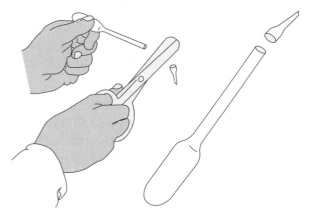

3. Cut a piece of masking tape about 6 cm long. Place it on a surface—such as a desk—on which you will be able to pull the tape back off without ruining the tape. Place the metric ruler next to it and transfer the cm and ½ cm markings to the tape. Your piece of tape should look similar to the one in **Figure 6.2**. Wrap the tape around the pipette with the 0 cm mark at the tip, or as close to the tip as possible. (See **Figure 6.2**.)

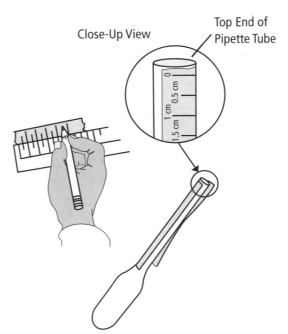

Close-Up View

Top End of Pipette Tube

4. Add 400 ml of water to the beaker; add a drop or two of food coloring to the water.

5. Add 12 BBs into the pipette so that all the BBs rest in the bulb of the pipette.

6. Lower the bulb end of the pipette into the colored water and release it. The bulb of the pipette should remain under water, and the open end should float almost vertically in the water.

7. Add BBs to the pipette one at a time while it is floating in the colored water. Stop adding BBs when only 1 or 2 cm of the pipette tube remains above the water. Remove the pipette from the water.

8. Plug the open end with a small piece of modeling clay as shown in **Figure 6.3**.

9. Place the pipette back into the colored water. If it floats so that the surface of the water is anywhere between the 0.5 cm and 2 cm marks, your hydrometer is complete. (See **Figure 6.4**.)

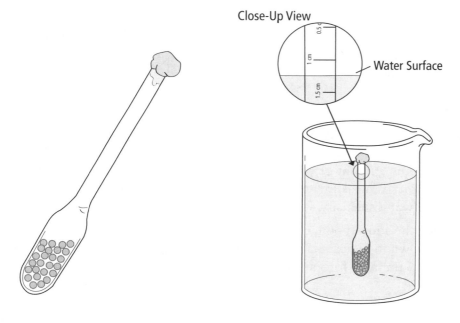

Close-Up View

Water Surface

Figure 6.3 (far left)
Diagram of the finished hydrometer with the BBs in the bulb and a modeling clay stopper in the tip

Figure 6.4 (right)
Hydrometer floating in the liquid. Note that the close-up view shows the water level at 1.25 cm below the tip.

10. If the hydrometer does not float properly, remove the clay plug and the BBs; go back to step 5 and proceed as directed. Once your hydrometer floats as described, you can use it to measure how the density of water changes as you add salt.

11. Place the hydrometer, bulb end down, in the colored water created in step 4. Wait until it stops bobbing up and down.

12. Look at the numbered lines on the tube of the hydrometer. Determine the number of the line that rests at the surface of the colored water. Write this number in the data table for Part 1 (**BLM 6.1**). If the hydrometer floats so that the water level is between two numbered lines, estimate (using decimal notation) the level where the tube touches the surface of the water. For example, the hydrometer in **Figure 6.4** shows a reading of 1.25 cm.

Activity 6

13. Measure three teaspoons or one tablespoon (15 ml) of pickling salt and pour it into the colored water. Using the hydrometer as a stirring rod, gently stir the water until all the salt has dissolved. Be careful not to disturb the clay plug.

14. Wait until the hydrometer stops bobbing up and down in the water. Determine the number of the line that is just touching the surface of the salty water in the same way you did in step 12. Write this number in the data table for Part 1 (**BLM 6.1**).

15. Add another three teaspoons or one tablespoon (15 ml) of dry pickling salt to the water. Stir as you did in step 13, and read the hydrometer when it stops moving. Record your observation for the "colored water + 30 ml salt" on the next line in the data table for Part 1 (**BLM 6.1**).

16. Complete the data table for Part 1 by adding salt to the colored water one teaspoon (15 ml) at a time as you did in the previous steps. Remember to stir, allowing for the water to settle before you make hydrometer readings.

17. Once you have completed all the readings on the different salt solutions, dump the solution into the waste container and return your materials to the teacher.

Part 2

1. Use the graduated cylinder to fill the jar or beaker about three-quarters full with fresh, cool water from the tap. In the data table for Part 2 (**BLM 6.2**), record the amount of water you place in the beaker; using your hydrometer from Part 1, measure and record the hydrometer reading, which is a reflection of the water's density. Remove your hydrometer from the beaker.

2. Using the soup spoon, carefully lower (do not drop!) the egg into the water (as shown in **Figure 6.5**) and observe what happens. Record your observations in the data table for Part 2 (**BLM 6.2**).

Materials for Part 2

Each group will need

- large jar or beaker
- pickling salt
- soupspoon (large enough to hold an egg)
- 100 ml graduated cylinder
- hard-boiled egg
- ruler or straightedge
- 5 ml metric measuring spoon (a teaspoon)
- hydrometer (from Part 1)

Time

45–50 minutes

Figure 6.5
Carefully lower the egg into the water.

3. Remove the egg from the water. Use the measuring spoon to add 10 ml of salt. (Your teacher will demonstrate how to level the salt in the teaspoon.) Add the salt to the water and stir thoroughly with your soupspoon. Measure the hydrometer reading of the solution and record it in the data table for Part 2 (**BLM 6.2**). After you remove the hydrometer, carefully put the egg back into the beaker. Record your observations in the data table for Part 2 (**BLM 6.2**).

4. Repeat step 3 until a change occurs.

Questions and Conclusions

1. The more salt that is dissolved in a solution, the _____ [higher / lower] the hydrometer floats.

2. A hydrometer reading of 4.5 cm would mean that _____ [no salt / some salt / lots of salt] is dissolved in a solution.

3. If ocean water gives a reading of 2.5 cm on your hydrometer, what reading might you get if you test freshwater? _____ Higher? Lower? The same?

4. Describe how the addition of salt affects some of the properties of water. Is the density of the water increasing or decreasing as salt is added to the water? How do you know whether the change is an increase or a decrease?

5. If an oceangoing ship is loaded with cargo so that 15 m of the ship rides below the surface of the water, what would you expect to happen when the ship enters the Mississippi River, which is only 16 m deep at some points?

6. What happened when the egg was placed in the freshwater?

7. Discuss the differences you saw when you put the egg in salt water compared to when you put it in freshwater.

8. How many milliliters of salt did you have to add to make the egg behave differently? Why is it important to use a straightedge to level the salt?

9. Why do you think things float better in salt water than in freshwater? What does the addition of salt do to the properties of the water?

10. What do you think might happen to the egg if it were left in the beaker for 24 hours? Two days? A week?

11. Compile your results with those of the rest of the class. Based on the class results, calculate the average amount of salt per 1,000 ml of water required to float the egg. Predict how much salt would be needed to float the egg in 20 L of water and in 300 L of water. Show your math.

12. Discuss how much the class results varied. What are some possible reasons for why they varied?

13. Why is it easier to float in the ocean than in a lake or pool?

What Can I Do?

You can measure salinity of natural waters and you might find fluctuations worth exploring. You can develop your own consistent method of sampling and measuring, or you can use existing methods. One source of instruction is the Environmental Protection Agency's Volunteer Estuary Monitoring Program. Search for "EPA Salinity" at *http://water.epa.gov/type/oceb/nep/monitor_index.cfm*. Another source is the GLOBE Program. Search for "GLOBE Salinity" at *http://globe.gov*.

Data Table: Part 1

Solution Being Tested	Hydrometer Reading
Colored Water	
Colored Water + 15 ml Salt	
Colored Water + 30 ml Salt	
Colored Water + 45 ml Salt	
Colored Water + 60 ml Salt	

Activity 6: Won't You BB My Hydrometer?

Data Table: Part 2		Amount of Water: _____ml
Total ml of Salt	**Hydrometer Reading**	**Result/Observations**

Class Data Table: Part 2		
Group	ml of Water	ml of Salt to Make Egg Float
Class Averages		

Won't You BB My Hydrometer?

What Is Happening?

The difference between freshwater and ocean water that is most obvious to students is the fact that the ocean is salty. Students may also have some understanding of the fact that the addition of salt to water changes some of the properties of the water. One important change resulting from the addition of dissolved salts is an increase in the density of the water. Water that contains dissolved substances has a greater mass per unit volume (density) than pure water. If 3 g of salt are dissolved in 100 ml of freshwater, the resulting mixture is 3% heavier than freshwater. Salinity is the amount of salt dissolved in water and can be expressed as the number of grams of salt per kilogram of ocean water. For a constant temperature and depth, the greater the salinity of ocean water, the greater its density. Open ocean water has an average salinity of about 35.

Students may know that it is easier to float in the ocean than in a pool. Some students may have even had the experience of swimming in Great Salt

Objective

Understand the implications of water density by building and using a hydrometer to measure the densities of freshwater and saltwater samples.

Key Concept

I: Properties of water

Materials for Part 1

Each group will need

- plastic transfer pipette
- sharp scissors
- fine-tip permanent marking pen (low or no VOC)
- metric ruler
- 20 BBs
- 500 ml beaker
- masking tape
- modeling clay
- food coloring
- pickling salt
- waste container
- towels or rags for cleanup
- teaspoon (5 ml) or tablespoon (15 ml)

Time

45–50 minutes

How Do We Know This?

The salt content of seawater is called salinity. How do research scientists measure salinity?

Originally, scientists measured salinity by evaporating the water and weighing the amount of salt left. The problem is that this method is quite inaccurate. In the latter 20th century, scientists began measuring ocean salinity by calculating the conductivity of seawater. The more salts in the water, the better it conducts electricity. This method is very accurate and has been used as the standard for decades.

Currently, two salinity satellite missions (*Aquarius* from NASA and *SMOS* from the European Space Agency) base their estimates of salinity by measuring the radiation that the ocean emits. Salinity affects the way radiation is emitted by seawater, and the satellites can measure this with remarkable accuracy. The objective is to measure salinity with an accuracy of 0.2 units. To give an idea of how much that is, consider that 0.2 salinity units corresponds to 1 g (about ¼ teaspoon) of salt per 5 L of water. These satellites can "taste" this salt from 600 km up in space.

Materials for Part 2

Each group will need

- large jar or beaker
- pickling salt
- soupspoon (large enough to hold an egg)
- 100 ml graduated cylinder
- hard-boiled egg
- ruler or straightedge
- 5 ml metric measuring spoon (a teaspoon)
- hydrometer (from Part 1)

Time

45–50 minutes

Lake in Utah, or seen images of the Dead Sea where floating in the water is almost effortless. The more salt dissolved in the water, the less water must be displaced by a floating object. Thus the denser the water, the higher (compared to the water level) an object will float. In this Activity, students investigate changes in the properties of water by building and using a hydrometer to measure relative density as increasing amounts of salt are dissolved in the water. Ocean salinity and ocean layers are covered in Activities 7 and 8.

The first part of this Activity is an engineering exercise requiring students to read and follow directions to build a hydrometer. Scientists use these simple devices to measure the relative density of liquids. In the second part of this Activity, students use their hydrometers to track the addition of salts to alter the density of water. At the beginning, the egg has a higher density than the freshwater and thus sinks when placed in the beaker. With the addition of the salt to the water, the water's density increases. When it has a higher density than the egg, the egg floats to the surface. It is possible for the density of the solution and the egg to be the same, in which case the egg will float below the surface of the water. The density of the egg remains constant throughout this Activity, while the density of the water increases as students add salt.

An Activity such as this, with a relatively straightforward procedure, encourages students to follow the principles of good scientific research. Students can control certain variables and are able to make predictions (by recording the amount of water used, by standardizing the amount of salt added, by standardizing the temperature of the water, and by comparing and combining their results with other groups in the classroom). Encourage students to not just follow the directions, but to think about why they do each step.

Preconceptions

This Activity is all about density, which can baffle students who are not comfortable with ratios. If students have done Activity 3, or other Activities in which they have physically explored relative densities, ask them to describe density in words, not just as an equation. If they have not done those Activities, ask them about dropping sticks and rocks in water. What happens? How do they explain it? Students may have the following preconceptions:

- All liquids have the same density.
- Density depends on the amount of liquid present.
- Weight determines if an object will sink or float.
- The amount of water will affect the way objects float or sink.
- Objects with holes sink.
- Objects with air float.

What Students Need to Understand

- The density of water increases with increasing salinity.
- The density of any liquid, not just a salt solution, can be measured using a calibrated float called a hydrometer.
- Hydrometers measure *relative* densities between different liquids. Freshwater is usually the standard for comparison.
- Substances of higher density will sink in substances of lower density, and substances of lower density will float in a substance of higher density.

Time Management

Each part of this Activity will require 45 to 50 minutes. Carefully store the hydrometers overnight.

Preparation and Procedure – Part 1

Demonstrate how to use a ruler or straightedge to level the salt in the spoon before adding it to the water. It is possible to use a clear plastic straw instead of a plastic transfer pipette simply by plugging both ends with a piece of clay. Be sure to have students label their hydrometers to avoid confusion on the second day of the Activity.

Preparation and Procedure – Part 2

Pickling salt is recommended because it is pure sodium chloride without additives. While students are working through the procedure, draw a table similar to the data table in **BLM 6.3** on the board. Once students have completed their tables, use their data to compile a class table and averages. Have students compare their results to those of other classmates and to the class averages.

Extended Learning

- Encourage students to take the hydrometers home and measure the densities of other common household fluids such as dishwashing liquid, milk, diet and regular soft drinks, cooking oil, real ocean water, water from a nearby creek, and so on. Have them keep a record of their findings and compare them with other students in the classroom. You may also want to challenge students to devise a way to use the hydrometer to give readings of salt concentrations in grams per liter, in parts per thousand, and in units of their own choosing.
- You may wish to have students try to float an egg in the middle of the beaker, rather than at the top. This can be accomplished by first adding sufficient salt to

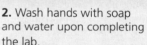

SAFETY ALERT

1. Be careful to quickly wipe up any spilled water on the floor—slip and fall hazard.

2. Wash hands with soap and water upon completing the lab.

float the egg, as in this Activity, then slowly adding freshwater to the beaker. Carefully adding freshwater will result in two layers of water with the egg suspended between them.

- Explore salinity measured automatically on buoys in bays and offshore: Are there patterns over time or across an area? The National Data Buoy Center provides real-time data online at *www.ndbc.noaa.gov*. As well, there are coastal observing systems, such as the Southern California Ocean Observing System (SCOOS) at *www.sccoos.org*.
- Have the class explore the use of hydrometers in making juice, or how hydrometers are used to measure the specific gravity of various petroleum products.

Connections

The oceans are a soup of ions and dissolved gases delivered by rivers and the atmosphere. Have students investigate some of the sources of ions and gases: weathering of rock and soil, undersea volcanic activity, and the solubility of gases (such as carbon dioxide) from the atmosphere, at *http://ga.water.usgs.gov/edu/whyoceansalty.html*.

Interdisciplinary Study

- Have students investigate the various uses of hydrometers. For example, hydrometers are used by auto mechanics to test radiator coolant. You may want to invite a local automobile service person to demonstrate the type(s) of hydrometer used by auto mechanics. Hydrometers are also often used in reef tank keeping.
- Students can study how animals adapt to changing salinity and density of water. Spotted sea trout, for instance, spawn where salinity varies, and their eggs change size and mass in response. Search the internet for terms such as *variable salinity sea trout eggs*.

Differentiated Learning

Students might enjoy looking at salinity variations throughout a region using Google Earth, a simple Geographic Information System (GIS). Google Earth is available for download at no cost. Students can view data from the National Data Buoy Center in Google Earth simply by selecting the command "Get Observations by…as KML."

Answers to Student Questions

1. The more salt that is dissolved in a solution, the *higher* the hydrometer floats.
2. A hydrometer reading of 4.5 cm would mean that *lots of salt* is dissolved in a solution.
3. If ocean water gives a reading of 2.5 cm on your hydrometer, you will get a lower reading if you test freshwater.
4. The addition of salt to the water increases the density of the water, making the hydrometer float higher in the water. The addition of salts to water can also affect the boiling and freezing points of water.

5. The ship would float lower in the freshwater of the Mississippi River than in the salty water of the Gulf of Mexico. This could cause the ship to run aground. Perhaps you could ask students what they think would happen if the ship were traveling through other liquids. Have students investigate what technologies freighters and other oceangoing ships use to negotiate movement between bodies of freshwater and salt water.

6. The egg sank in the freshwater.

7. After a sufficient amount of salt was added to the freshwater, the egg floated.

8. Answers will vary. Since the results are going to be compared to the rest of the class, the leveling ensures that everyone adds the same amount of salt. This precision is an important part of good scientific research.

9. Salt water has a higher density than freshwater, and thus things will more likely float in the salt water than in freshwater. The addition of salt to water increases its density.

10. Answers will vary. Have students try this to test their hypotheses.

11. Answers will vary. Make sure that students make the appropriate conversions between milliliters and liters when calculating the amount of salt required to float an egg in 20 L or 300 L of water.

12. Answers will vary. Results might vary because of inaccuracy inherent to measurements, variation in the eggs, incomplete dissolution of the salt, miscalculations, and other reasons.

13. The density of the water in the ocean is higher than that in a freshwater lake or pool, thus making it easier for a person to float.

Assessment

- Circulate through the lab to help students as they do this Activity. Ask questions such as, "What are you observing?" "Can you make any predictions about what you'll see when you add the next batch of salt?" "Why do you predict that?"
- As a summative assessment, you could ask students to write a brief description of what they did, the results, and what they learned. You can ask them how they would change their methods, especially after reviewing and discussing results from the entire class.
- You can also grade students' answers to the questions.

Resources

www.ndbc.noaa.gov

www.sccoos.org

http://ga.water.usgs.gov/edu/whyoceansalty.html

Activity 7 Summary

Students add three layers of different salinity and color of water to a straw to create layers based on the solutions' densities. Students' observations allow them to understand salinity and density stratification in estuaries, bays, and oceans, and the mixing that occurs at layer boundaries.

Activity	Subject and Content	Objective	Materials
Ocean Layers	Salinity/density layers in water	Investigate what happens when ocean water, brackish water, and river water contact one another.	Each group will need: cafeteria tray, slice of clay 3 cm thick, clear plastic straw (about 10 cm long), three 250 ml clear plastic cups containing 25 ml each of the colored solutions, 250 ml clear plastic cup (waste container for used solutions), three medicine droppers or plastic pipettes, one or two sheets of white paper, towels or rags for cleanup

Time	Vocabulary	Key Concept	Margin Features
50 minutes	Estuary, Brackish water	II: Ocean structure and water movement	Safety Alert!, Fast Fact, What Can I Do?, Connections, Resources

Scientific Inquiry	Unifying Concepts and Processes
Modeling and observing	Change within open systems

Ocean Layers

Background

Ocean water is not the same everywhere. In some places, the water is colder or deeper than in other places. Some parts are denser or contain differing amounts of dissolved salts than other parts. All these things affect the way ocean water behaves.

Although water is the most abundant substance on Earth's surface, very little of it is pure water. Many elements other than hydrogen and oxygen—the only two elements in pure water—are found in Earth's water. Tap water, for example, contains chemicals used to disinfect the water and to prevent bacterial growth. Ocean water has many other elements in it, such as dissolved salts and other materials.

Water has different properties depending on the environment in which it is found. Water in streams and rivers has few dissolved salts and is called freshwater. Water found in the ocean is called salt water because it contains a lot of dissolved salt. In areas where rivers flow into oceans—called **estuaries**—**brackish water** is found. Brackish means that the water has a higher concentration of salts than river water, but a lower concentration of salts than the open ocean.

In this Activity, exploring how these three types of water—salt, brackish, and fresh—mix will help you understand what effects this mixing has on Earth's environment.

Vocabulary

Estuary: A semienclosed body of water, most often near the mouth of a river or other freshwater source, where freshwater and salt water mix.

Brackish water: The type of water found where freshwater and salt water mix. The salt content of the water usually varies, and the water is considered neither fresh nor salty.

Topic: properties of ocean water
Go to: *www.scilinks.org*
Code: PESO 004

Topic: ocean water chemistry
Go to: *www.scilinks.org*
Code: PESO 005

Objective
Investigate what happens when ocean water, brackish water, and river water contact one another.

Activity 7

Materials

Each group will need
- cafeteria tray
- slice of clay 3 cm thick
- clear plastic straw (about 10 cm long)
- three 250 ml clear plastic cups containing 25 ml each of the colored solutions
- 250 ml clear plastic cup (waste container for used solutions)
- three medicine droppers or plastic pipettes
- one or two sheets of white paper
- towels or rags for cleanup

Time

50 minutes

Figure 7.1
Diagram showing the correct placement of the straw in the modeling clay

SAFETY ALERT

1. Be careful to quickly wipe up any spilled water on the floor—slip and fall hazard.

2. Wash hands with soap and water upon completing the lab.

Procedure

1. Have a member from your group pick up a tray of materials. Each tray contains three different solutions with different salinities (and therefore different densities) and colors so you can tell them apart.

2. Stick the plastic straw into the slice of clay at an angle as shown in **Figure 7.1**. (Note: Do not stick the straw all the way through the clay. If the straw comes out through the bottom of the slice of clay, remove the straw and stick it into a different place on the slice.)

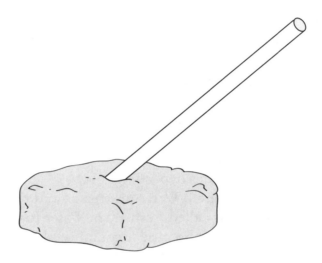

3. Test the straw for leaks by filling it with tap water. If the water leaks out of the bottom of the straw, remove the straw and stick it into a different spot on the clay. Test for leaks again.

4. When you are sure that the straw is not leaking, empty the water into your waste container by picking up the entire assembly (the block of clay and the straw) and tipping it so the water drains out of the straw. You may need to shake the straw several times to get all of the water out, but do not remove the straw from the clay.

5. You know only these two things about the three solutions on your tray:

   (a) The only difference among the three solutions is their color and the amount of salt they contain.

   (b) If you add small amounts of each solution to the straw in the correct order, you will produce three distinct layers that do not mix.

   Your goal is to explore how the salt water, brackish water, and freshwater interact by adding a bit of each solution to the straw as described in step 6. However, first state the order in which you want to add the liquids and explain your reasoning.

6. Add a small amount of each solution to the straw. **Figure 7.2** illustrates how to do this using a medicine dropper. Fill about one-third of the straw with each solution. Place a white piece of paper behind the straw and observe the colored solutions inside. Record your observations in the data table in **BLM 7.1**.

Fast Fact
During World War II, German submarines (U-boats) are reputed to have used salinity currents to sneak silently into the Mediterranean Sea. Very salty water flows out of the Straits of Gibraltar in a deep layer, while less salty water flows from the Atlantic into the Mediterranean as a surface layer. U-boats drifted silently past Allied listening posts under water using the surface currents.

Figure 7.2
Add a small amount of each solution to the straw with the dropper.

7. If you saw three distinct layers, state your conclusions about the relative saltiness of the three layers. If you did *not* see three distinct layers, write a hypothesis about the relative saltiness of each of the colored solutions.

8. Empty the contents of the straw and revise the order in which you fill the straw to test your hypothesis. Record your observations.

9. If you did not see three distinct layers, continue revising your hypothesis and testing it. Record new hypotheses and observations in the data table (**BLM 7.1**).

Questions and Conclusions

1. What was the order of adding solutions that resulted in three distinct layers?

2. How can you explain this result?

3. Suppose you are at the seacoast in an area where a river runs into a some-what salty bay before it reaches the ocean. Where do you predict you would find the *saltiest* water: near the surface of the bay or near its bottom? Explain why.

What Can I Do?

Try doing this Activity with a friend or someone in your family. If you made a hydrometer in Activity 6, measure the salinity of the water in each layer. If you live near salt water and swim after a heavy rain, see if you can detect layers—with your senses or by collecting samples at different depths and using a hydrometer. (If the layers have different temperatures, let the samples sit until they are the same, then measure the salinity.)

Data Table: Hypothesis and Observations

Experiment 1: _____

_____

Observation: _____

_____

_____

Experiment 2: _____

_____

Observation: _____

_____

_____

Experiment 3: _____

_____

Observation: _____

_____

_____

Experiment 4: _____

_____

Observation: _____

_____

_____

Ocean Layers

What Is Happening?

Water is the most abundant chemical on the surface of Earth. Since so many substances dissolve in it, liquid water is almost never found in a pure form, except in laboratories. Ocean water, for example, is a complex solution of about 96% pure water, 3.0% NaCl (common salt), and smaller amounts of many other chemicals, most of which are salts, and some gases (see **Table 7.1**).

The amount of dissolved salts is not constant from place to place or from time to time throughout the ocean. As a consequence, different water masses within the ocean can have very different properties. The density of ocean water, for example, is strongly affected by water temperature and amount of dissolved salts. As a result, density varies from place to place. As temperature decreases, density increases until it reaches a maximum at 4°C in freshwater. Below this temperature, the density begins to decrease. (Remember this from Activity 3.) Additionally, as more salts dissolve in water, the water becomes denser. This trend can continue until the water becomes saturated with salts. Each factor contributes to the variation in ocean water from one region to the next.

One obvious consequence of differing densities in ocean water is layering, which results in fairly distinct layers of ocean water as depth increases. A less obvious consequence is the formation of density currents. At Earth's poles, the water is both very salty (because during the formation of ice, the salts in seawater are left in the liquid part) and very cold. These conditions make for very dense water. This water sinks and flows toward the equator along the

How Do We Know This?

How do we know that estuaries and oceans are layered?

The oceans are layered in some cases because densities differ among water masses. The densest water mass will be on the bottom, and the least dense water mass will be on the surface. Seawater density is a result of temperature and salinity (lowest temperature and highest salinity correspond to highest density). Traditionally, seawater density, and hence ocean layers, are measured using a CTD, which stands for conductivity (a measure of salinity), temperature, and depth. More modern methods consist of detecting density interfaces between two layers using a high-frequency signal generated from a boat, much the same as a fish finder.

Objective
Investigate what happens when ocean water, brackish water, and river water contact one another.

Key Concept
III: Ocean structure and water movement

Materials
Each group will need
- cafeteria tray
- slice of clay 3 cm thick
- clear plastic straw (about 10 cm long)
- three 250 ml clear plastic cups containing 25 ml each of the colored solutions
- 250 ml clear plastic cup (waste container for used solutions)
- three medicine droppers or plastic pipettes
- one or two sheets of white paper
- towels or rags for cleanup

Time
50 minutes

ocean bottom. Water at the equator flows toward the poles near the surface to replace the sinking water, contributing to global circulation patterns. These circulation patterns are also very important for regulating Earth's climate.

Typically, students are aware that different kinds of water have differing amounts of salt dissolved in them. River water, or freshwater, has only a small amount of dissolved salt. Areas where rivers enter the ocean—called estuaries—contain brackish water and have a higher salt content than freshwater rivers. Ocean water contains the highest salt content.

Colored solutions are used in this Activity to represent three common types of water: ocean (salt) water, brackish water, and river (fresh) water. The amount of salt dissolved in each is one, but by no means the only, difference among these three kinds of water. The different concentrations of salt lead to differing densities among these types of water. Fluids having different densities tend to form layers.

Since ocean water contains large amounts of NaCl and other dissolved salts, ocean water is denser than river water. Under certain conditions, river water flowing into brackish estuaries can form a separate layer on top of the denser ocean water.

Currents, wind conditions, water temperature, and other factors—some of which will be explored in later Activities—affect the ways freshwater and ocean water mix or form discrete layers. In this Activity, students will discover the effects of the amounts of dissolved salts on water density and layering. Layering and mixing in estuaries are discussed in more detail in Activity 9.

Table 7.1: Composition of Ocean Water

Element	Weight Percent of Ocean Water
Oxygen	85.7000
Hydrogen	10.8000
Chlorine	1.9000
Sodium	1.0500
Magnesium	0.1350
Sulfur	0.0885
Calcium	0.0400
Potassium	0.0380
Bromine	0.0065
Carbon	0.0028

The following elements are found in seawater at less than 0.001%, and are listed in decreasing order from Strontium at 0.00081% to Radon at 0.0000000000000006%: Strontium, Boron, Silicon, Fluorine, Argon, Nitrogen, Lithium, Rubidium, Phosphorus, Iodine, Barium, Aluminum, Iron, Indium, Molybdenum, Zinc, Nickel, Arsenic, Copper, Tin, Uranium, Krypton, Manganese, Vanadium, Titanium, Cesium, Cerium, Antimony, Silver, Yttrium, Cobalt, Neon, Cadmium, Tungsten, Selenium, Germanium, Xenon, Chromium, Thorium, Gallium, Mercury, Lead, Zirconium, Bismuth, Lanthanum, Gold, Niobium, Thallium, Hafnium, Helium, Tantalum, Beryllium, Protactinium, Radium, Radon.

Preconceptions

Ask students to discuss in their groups or as a Think-Pair-Share how they think salinity affects density and what their evidence or reasoning is. The following are some possible preconceptions students might have:

- Ocean water salinity remains the same, regardless of location or depth.
- The ocean is vertically homogeneous.
- Oceanic currents are the same on the seafloor as they are on the surface.

What Students Need to Understand

- The colored solutions are models for demonstrating how river water, brackish water, and ocean water *mix* or form layers. Water from natural sources has a greater range of dissolved substances than the solutions in this Activity.
- When layers form in the ocean, the water (assuming equal temperature) containing the most dissolved salt tends to form the bottom layer. The water having the least amount of salt is usually on top, because it is the least dense.
- As more solids are dissolved in water, the water becomes more dense.
- The density difference between two volumes of water can prevent them from easily mixing. Less dense water will remain on top of denser water.

Time Management

This Activity can be completed in 50 minutes, but be sure to leave at least 5 minutes for students to clean up. Also, if you choose to discuss the last question in particular as a class, allow 5 to 10 minutes for that. The question could also be a good way to engage students at the beginning of class the next day.

Preparation and Procedure

The pickling salt recommended for this Activity produces very clear salt solutions. Other types of salt may produce somewhat cloudy solutions because of various substances added to them. Clear solutions make it easier to determine whether different samples are mixing or forming layers.

If you use a different kind of salt, test the proportions of salt and water before class to be sure that they give clear separations of the layers. As a guideline, about the same volume of table salt can be substituted for the pickling salt. If you use coarser salt (such as kosher salt or ice cream salt), you may get better layers if you increase the amount of salt in the "ocean water" to about 120 ml in 500 ml water, and in "brackish water" to about 40 ml in 500 ml water.

> **SAFETY ALERT**
>
> **1.** Be careful to quickly wipe up any spilled water on the floor—slip and fall hazard.
>
> **2.** Wash hands with soap and water upon completing the lab.

Have students assist in preparing the salt solutions. This way they can see how the solutions are prepared and understand the differences among them. It will reduce their tendency to use extraneous facts, such as the color of the water, to explain the events observed.

Prepare the colored solutions as follows:

Ocean Water	Brackish Water	River Water
500 ml water	500 ml water	500 ml water
90 ml salt	30 ml salt	no salt
20 drops blue food coloring	20 drops red food coloring	20 drops green food coloring

Stir the "ocean water" and the "brackish water" until the salt is totally dissolved. Place each solution in a jar or beaker with no labeling that indicates the contents.

Extended Learning

• Challenge students to discover the smallest amount of salt that they can dissolve in 500 ml water that will still allow a layer to form between the "ocean water" and pure water.

• What is the maximum number of discrete layers of colored salt water that students can place in a straw? In order to answer this, students may wish to experiment with different methods of forming the layers, and test what minimum difference in salt concentration is necessary to form a layer.

• Encourage students to investigate the different layers of water in the ocean, which occur because of density differences. They may also find it interesting to research how long it takes a particle of water to travel in a density current from the poles to the equator (see Activity 8).

• If your school is near an estuary, you can collect water samples from different depths and measure the salinity. Use a hydrometer, salinity meter, salinity test kit, or refractometer. (See How Do We Know This?)

Interdisciplinary Learning

• Have students explore layering of water in estuaries. The layering produces a dynamic ecosystem with diverse species of animal life. You can ask students to investigate the zonation of attached ("sessile") organisms—how different organisms are adapted to living in water layers of different salinity and temperature. For example, do barnacle or seaweed species vary depending on depth?

• Bottled water offers interesting comparisons. Some bottled water manufacturers add different salts to get a particular taste. You could have the class discuss mineral water versus distilled water. Mineral waters from different parts of the world also offer a point of comparison and discussion among students.

Differentiated Learning

- Once students have sorted the water into layers by salinity—and density—you can ask them to measure the density using hydrometers from Activity 6. Students in advanced mathematics can graph the amount of salt added versus the density of the solution. They can also try to determine the algebraic relationship between them.

- Students who like designing or building things can design and build a water sampler. Cornell University's Environmental Inquiry Engineering Design Challenge suggests ways to set up a design problem. Search for "stormwater design challenge" at *http://ei.cornell.edu/watersheds/challenges*. You could ask students to draw their device with free, downloadable 3-D drawing software available at *http://sketchup.google.com*.

Connections

Just as an estuary has layers, so does the atmosphere. Ask students to watch for examples of air rising and then reaching a certain level and spreading out. Examples might include smoke, the "anvil" at the top of a thunderstorm, or stratus clouds.

Answers to Student Questions

1. The order of adding solutions that resulted in three distinct layers was, from bottom to top: blue (ocean or salt water); red (brackish water); green (river water or freshwater).

2. The blue water has the highest density, the green water has the lowest density, and the density of the red water is between the other two. Liquids of different densities will form layers based on their respective densities.

3. Salt water has a higher density than freshwater; therefore, it would be found near the bottom of the bay. The freshwater from the river would be found near the surface of the bay. Liquids of different densities will form layers based on their respective densities.

Resources

http://ei.cornell.edu/ watersheds/challenges/

http://sketchup.google.com

Assessment

- For formative assessment during the Activity, circulate and ask students about their observations, hypotheses, and next steps. Ask them, "If you were to do this again, how would you approach the problem?"

- For a performance summative assessment, give students three new solutions with the same salinities as before, but with different colors. Do not add the food coloring in front of students. Ask them to figure out which color represents the ocean water, the brackish water, and the freshwater.

- You could ask students, "Please explain what you know now about how salinity affects density. What do you predict will happen when a heavy rainstorm falls on land? On a bay?"

- You could also grade students' answers to questions.

Activity 8 Planner

Activity 8 Summary

Students explore the science behind the ancient story of bodies suspended in limbo in the ocean. Students float weighted vials or test tubes within a column of water made by layering cold salt water, cold freshwater, and hot freshwater. Density differences in seawater cause different ocean layers as well as ocean currents.

Activity	Subject and Content	Objective	Materials
The Myth of Davy Jones's Locker	Density layering of ocean	Investigate some of the properties of water that could explain the myth of Davy Jones's Locker.	For the demonstration, the teacher will need: three buckets (4 L or 1 gal. size), eight small screw-top vials (or eight test tubes and stoppers), plastic cylinder with an end cap (122 cm [4 ft.] tall and 4 cm [1.5 in.] in diameter), ring stand with clamp to hold cylinder, large funnel, 1 to 2 m length of rubber tubing with U-shaped glass tubing in one end, package of BBs, hot plate and pan or coffee heater, pickling salt (produces a clear brine), ice, water, red wax marking pencil, optional: food coloring
			For the Activity, each student or group will need: three 600 ml beakers, eight small screw-top vials (or eight test tubes and stoppers), plastic cylinder with an end cap (122 cm [4 ft.] tall and 4 cm [1.5 in.] in diameter), ring stand with clamp to hold cylinder, large funnel, 0.5 m length of rubber tubing with U-shaped glass tubing in one end, package of BBs, pickling salt, ice, water, red wax marking pencil

Time	Key Concepts	Margin Features
(Demonstration Preparation): 25–30 minutes (Activity): 50 minutes	I: Properties of water II: Ocean structure and water movement	Safety Alert!, Fast Fact, What Can I Do?, Connections

Scientific Inquiry	Unifying Concepts and Processes	Personal/ Social Perspectives	Historical Context
Modeling and observing	Effect of density on ocean's vertical structure	Historical perspectives of Earth	Seafaring mythology

The Myth of Davy Jones's Locker

Background

For centuries, sailors believed that bodies buried or lost at sea did not sink to the bottom. They believed that a special depth existed between the surface and the bottom of the ocean where a body would remain suspended. Sailors called this region of the sea "Davy Jones's Locker."

Your teacher has set up a large column in the classroom with vials floating at various levels throughout the column. These vials exhibit different buoyancies. Study the column carefully and, based on what you have learned in previous Activities, develop some hypotheses that could explain this phenomenon. In this Activity, you will create your own model of Davy Jones's Locker.

The results of this experiment might lead you to believe that Davy Jones's Locker could exist. However, this experiment is entitled The Myth of Davy Jones's Locker. As is often the case, experiments conducted in the classroom do not mirror the complexities of the real world. Davy Jones's Locker does not really exist.

When bodies fall into the ocean and sink, the pressure of the overlying water increases to tremendous levels—far more than could be modeled in the classroom. Increased pressures eventually will crush the body, and it will sink to the bottom.

This Activity does, however, highlight some important aspects of how layers form according to density, called density layering. The ocean exhibits a number of layers, and this layering is important in understanding estuary formation (which will be examined in Activity 9) and deep ocean currents.

Topic: density
Go to: *www.scilinks.org*
Code: PESO 006

Topic: properties of ocean water
Go to: *www.scilinks.org*
Code: PESO 004

Objective

Investigate some of the properties of water that could explain the myth of Davy Jones's Locker.

Activity 8

Procedure

Part 1: Observation

Examine the demonstration column carefully. In the data table in **BLM 8.1**, record at least five observations about the column that you think are relevant to explaining this phenomenon.

Part 2: Hypothesis

Based on your observations from previous Activities that you have completed in oceanography, write a hypothesis in the data table (**BLM 8.1**) that you believe explains the phenomenon observed.

Part 3: Prepare Your Own Column

1. Label three beakers and fill them according to the following guidelines:

   Beaker 1—Ice Cold Ocean Water: Fill the beaker with tap water, adding ice to chill. Stir pickling salt into the beaker until no more will dissolve. (The salinity of ocean water averages 35 g of salt per liter of water. However, the more salt that dissolves in your "ocean water" solution, the easier it will be to set up the water column.)

   Beaker 2—Cold Tap Water: Fill the beaker with tap water. You may need to add some ice to chill the water slightly, depending on the temperature of the tap water.

   Beaker 3—Hot Tap Water: Fill the beaker with hot tap water. This water should be as hot as possible, but not so hot that it will scald someone if spilled.

2. Add varying numbers of BBs to eight screw-top vials until:
   - two vials sink in Beaker 1 (cold ocean water)—label these vials A
   - two vials sink in Beaker 2 (cold tap water) BUT float in Beaker 1—label these vials B
   - two vials sink in Beaker 3 (warm tap water) BUT float in Beaker 2—label these vials C
   - two vials float in Beaker 3—label these vials D

   Note: It is important to test vials B and C carefully to make sure that they float and sink appropriately. Also, be careful to cap all vials tightly to prevent leaks.

3. Cap the bottom of the cylinder and place the vials into the empty cylinder. Using a ring stand and clamp, secure the tube so it stands upright (vertical).

4. Discuss with your group in which order to fill the cylinder to minimize mixing. When you agree, show the order to your teacher. To minimize mixing of the layers, use the device shown in **Figure 8.1**: a glass U-shaped tube attached to one end of a 0.5 m length of rubber tubing and a widemouthed funnel to fill the cylinder. Allow the layers to settle for a few seconds.

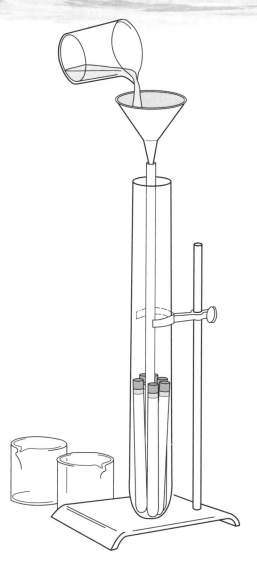

Fast Fact

Davy Jones's Locker appears in songs, movies, TV shows, plays, and books. It is mentioned in the U.S. Navy's song "Anchors Aweigh," *Pirates of the Caribbean*, *SpongeBob SquarePants*, *Peter and Wendy* (the play that is the basis of *Peter Pan*), and *Treasure Island*. Books by Charles Dickens, Herman Melville, Edgar Allen Poe, and others include it as well.

Figure 8.1
Carefully pour the water from beakers 1, 2, and 3 into the cylinder as illustrated.

What Can I Do?

Several hundred years ago, Italian scientist Galileo invented a type of thermometer based on the principle that matter becomes denser as it cools and less dense when it heats up. You can learn about how Galileo's thermometers work and even build one with the help of an adult. Search online for "homemade Galilean thermometer jar" for instructions to make a thermometer from baby food jars (or other small containers) and a large vase.

5. After adding the contents from each beaker, observe what happens. Record your observations in the data table (**BLM 8.1**).

Questions and Conclusions

1. What are the characteristics of each of the three layers in the column?

2. What do you think is responsible for this layering effect? Explain your response.

3. Evaluate your hypothesis from Part 2 of the procedure. Were you on the right track?

4. Predict how long these layers would remain distinct if left undisturbed.

5. List what you already know about layers in the ocean. Now evaluate the strengths and weaknesses of this model of oceanic layers.

Data Table: Hypothesis and Observations

Observations: _____

_____

_____

_____

_____

_____

Hypothesis: _____

_____

_____

Observations From Procedure Part 3 Step 5: _____

_____

_____

_____

_____

_____

The Myth of Davy Jones's Locker

What Is Happening?

Many variables can affect water's density—the amount of dissolved salts (salinity) and temperature are two common examples. Increasing salinity will increase the density of water; increasing temperature will decrease the density of water. In this Activity, the layer at the bottom of the cylinder is denser than the layers above it; therefore, it stays at the bottom. The top layer is the least dense and floats on the other two. Students should be able to identify salinity and temperature as the two factors that affect the density of water in this demonstration.

Salinity and temperature are important in explaining why the ocean is stratified into three separate layers—like the models you and students have constructed. In most regions of Earth's oceans, the surface, or mixed layer, contains the warmest and least dense waters in the ocean. The middle layer is a transitional area where water density changes markedly with depth. The

How Do We Know This?

How do we know the velocity of deep ocean currents? Surface currents seem easy to measure—you drop in a drifting buoy. What about deep-water ones though?

There are two ways in which scientists obtain information about deep ocean currents: direct observation and indirect estimation. One direct observation method is the release of drifters that travel at a specific depth or that follow water of a certain density. These drifters provide information either by coming to the surface and sending signals to passing satellites or by emitting acoustic pulses that are received at measuring stations near the coast. Another method of direct observation is by using moorings in certain locations with instruments called current meters that measure the velocity of the currents either by direct mechanical methods (like a weather vane) or by using the acoustic Doppler effect (change in wave frequency caused by relative motion). A method of indirect estimation of deep currents uses the fact that these currents are part of the thermohaline circulation; this implies that by knowing the water density, we can estimate the velocity of the interior of the ocean (away from the coast, the surface, and the bottom).

Objective

Investigate some of the properties of water that could explain the myth of Davy Jones's Locker.

Key Concepts

I: Properties of water
II: Ocean structure and movement

Materials (Demonstration)

Teacher will need
- three buckets (4 L or 1 gal. size)
- eight small screw-top vials (or eight test tubes and stoppers)
- plastic cylinder with an end cap (122 cm [4 ft.] tall and 4 cm [1.5 in.] in diameter)
- ring stand with clamp to hold cylinder
- large funnel
- 1 to 2 m length of rubber tubing with U-shaped glass tubing in one end
- package of BBs
- hot plate and pan or coffee heater
- pickling salt (produces a clear brine)
- ice
- water
- red wax marking pencil
- optional: food coloring

Time (Demonstration Preparation)

25–30 minutes

bottom, or deep, layer makes up approximately 80% of the ocean's volume and contains the coldest, saltiest and, therefore, most dense water in the ocean. While some mixing occurs between them, the layers persist and are well defined. (See **Figure 8.2**.)

Figure 8.2
Idealized diagram showing the surface "mixed" layer, the transitional layer, and the deep layer in relation to the configuration of the sea floor

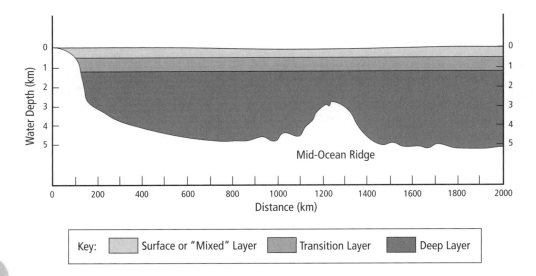

Key: Surface or "Mixed" Layer   Transition Layer   Deep Layer

Materials (Activity)

Each student or group will need
- three 600 ml beakers
- eight small screw-top vials (or eight test tubes and stoppers)
- plastic cylinder with an end cap (122 cm [4 ft.] tall and 4 cm [1.5 in.] in diameter)
- ring stand with clamp to hold cylinder
- large funnel
- 0.5 m length of rubber tubing with U-shaped glass tubing in one end
- package of BBs
- pickling salt
- water
- red wax marking pencil

Time (Activity)

50 minutes

The density differences between the layers and the pull of gravity on water of different densities cause deep ocean currents called thermohaline currents (*thermo* meaning temperature and *haline* meaning salt content). These density-driven currents begin in polar regions. At and near the poles, the water is cold—in fact, it can be below 0°C because the dissolved salts lower the freezing point of the water. This cold water is much denser than water found at lower latitudes. Also, water near the poles is continually being frozen into ice. Since this process freezes only the water, the salts are left behind causing the unfrozen ocean water to become even saltier. This combination of low temperature and high salinity causes water near the poles to be very dense and sink. Deep ocean currents are established as the dense polar water sinks to the bottom of the ocean and flows toward the equator. These currents move very slowly. Speeds of 1 to 2 cm per sec. are typical— that is about 0.023–0.045 mph. At this rate, it would take more than 15 years for the water in these currents to travel the roughly 10,000 km from the poles to the equator. By contrast, the flow rate of the main surface currents is more than 100 cm per sec.—at least 50 times the speed of some deep ocean currents.

In this Activity, density layering is reinforced and you can introduce students to the concept of deep ocean currents. You may want to refer to this Activity during Activity 10 to reinforce the distinction between surface and deepwater currents.

Preconceptions

Like Activities 4, 6, and 7, this Activity hinges on students understanding density. Ask students to review the concept of density by having them state what they know about it in their own style, drawing on specific examples from their previous experience. Here are possible preconceptions students might have:

- Density depends on how much of a substance is present.
- All ocean currents are driven by wind.

What Students Need to Understand

- Density of liquid water increases with decreasing temperature.
- Density of liquid water increases with increasing salinity.
- The water in the ocean is layered because of differences in densities of various masses of water.
- Differing densities of ocean water result in the movement of ocean water.

Time Management

The length of this Activity will vary depending on the choice of presentation. Demonstration, observation, experimentation, and discussion of this Activity may be completed in 50 minutes. Preparation for the demonstration will take approximately 25 to 30 minutes.

Preparation and Procedure

In this Activity, students will observe vials suspended at different points in a large demonstration column of water that you have already set up.

Repeating this demonstration another day, using food coloring to color the various layers, will clearly illustrate the relationship between the position of the vials and the different water layers.

The stratified layers of the water column will slowly mix together; however, the cylinder should retain most of its layering for a full day. To help maintain the layers, you may wish to place the lower section of the cylinder in a bucket of ice water between class periods and periodically add warm water to the top layer. If left in place over several days, the top layer will lose enough heat that a two-layer column will result—freshwater on the top and salt water on the bottom. Without agitation or mixing, these layers will remain distinct for quite some time.

SAFETY ALERT

1. Be careful to quickly wipe up any spilled water on the floor—slip and fall hazard.

2. Wash hands with soap and water upon completing the lab.

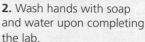

To Prepare the Demonstration

You will need to find or make a column for this demonstration. Clear 2½ in. PVC pipe and caps are available from specialty plumbing supply stores. You can order it online in 10 ft. lengths for about $12/ft. (2011 prices). Suppliers will cut it and ship it in 5 ft. lengths.

Set up the water column as follows: (Note: Students will go through the same steps to set up their columns, but will use different, much less expensive, equipment.)

Label three buckets and fill according to these guidelines:

1. **Bucket 1—Ice Cold Ocean Water:** Fill the bucket with tap water, adding ice to chill the water. Stir pickling salt into the bucket until no more will dissolve. (The salinity of ocean water averages 35 g of salt per liter of water. However, the more saturated your "ocean water" solution, the easier it will be to set up the water column.)

   Bucket 2—Cold Tap Water: Fill the bucket with tap water. You may need to add some ice to chill the water slightly, depending on the temperature of the tap water.

   Bucket 3—Warm Tap Water: Fill the bucket two-thirds full with tap water. Take the remaining water required to fill the bucket and heat it on a hot plate to just before boiling. Add the heated water to Bucket 3. This produces a warm tap water.

2. Add varying numbers of BBs to eight screw-top vials until:
   - two vials sink in Bucket 1 (cold ocean water)—label these vials A
   - two vials sink in Bucket 2 (cold tap water) BUT float in Bucket 1—label these vials B
   - two vials sink in Bucket 3 (warm tap water) BUT float in Bucket 2—1abel these vials C
   - two vials float in Bucket 3—label these vials D

   Note: It is very important to test vials B and C carefully to make sure they float and sink appropriately. Otherwise, vials may not float in the middle of the column. Also, be careful to cap all vials tightly to prevent leaks.

3. Cap the bottom of the cylinder and place the vials into it. Using a ring stand and clamp, secure the cylinder so it stands upright (vertical).

4. Carefully pour the water from Buckets 1, 2, and 3 into the cylinder in the following order: Bucket 1 first, Bucket 2 second, and Bucket 3 last. To minimize mixing of the layers, use the device shown in **Figure 8.3**—a glass U-shaped tube attached to one end of a 1 to 2 m length of rubber tubing and wide-mouthed funnel to the other end of the rubber tubing.

   Note: You may need some assistance from a student or another teacher when you get ready to pour the water into the demonstration tube.

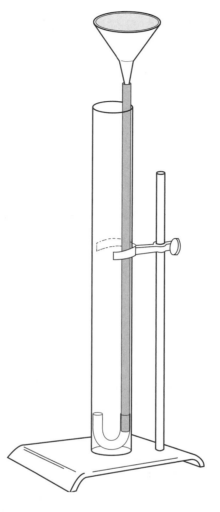

Figure 8.3
Demonstration cylinder with U-shaped tube attached

5. The vials should space themselves throughout the column corresponding to water densities—vial D at the surface, vial C at the boundary between the layers of warm and cold tap water, vial B at the boundary between the layers of cold tap water and cold salt water, and vial A at the bottom of the cylinder.

Cylinders for students to use are easier to find and less expensive. If you have 500 ml graduated cylinders, you can use them. An inexpensive alternative is to use 4 ft. Clear Tube Guards for fluorescent light bulbs from home construction stores. Choose the 1¼ in. ones. (They cost about $3.) Caps for 1¼ in. PVC pipes fit the ends tightly, for less than $1 each. You will want to caulk the caps on the cylinders to be watertight. These tubes will be fine to use for students' cylinders if you cut them in half.

You will want to leave at least one of the students' cylinders set up for several days for them to be able to answer question 4. Let them examine the cylinder daily to observe changes that will let them predict how long the layers will take to mix.

Connections

Unstable air moves upward through the atmosphere until its density is the same as the surrounding air—we see the effects as cumulous clouds. Ask students to investigate the idea of stable air in the troposphere, when lower levels are cooler and denser than higher levels, and unstable air, when the opposite occurs. What are some of the ways stable air becomes unstable? (e.g., heating from below on a sunny summer day or forest fires). What are some of the consequences of stable air? (e.g., layers of smog that will not dissipate)

Extended Learning

- To illustrate the temperature gradient in a column of ocean water, attach a string to a thermometer and lower it into the demonstration tube. Take temperature readings every 5 cm. Record and plot temperature readings on a graph, showing the change in temperature with increasing depth.
- Add some food coloring to tap water and then freeze the water into ice cubes. Place one of these colored cubes in the water column and observe the results. Ask students to explain what they are seeing.

Interdisciplinary Learning

- Have students study and research the myth of Davy Jones's Locker. When and how did this myth arise? Why did it seem reasonable to people at that time? How has it been used in literature and entertainment? (See Fast Fact for examples.)
- What are some other myths that early sailors and explorers told about the sea? How were these myths dispelled, if they have been? The Bermuda Triangle is a myth that persists today. Have students investigate what scientists or historians say about this and other myths.
- Have students investigate how submarines, marine mammals, and fish control their buoyancy as they dive thousands of feet into water of increasing density.

Differentiated Learning

For students who need practice with measuring density, you can ask them to use hydrometers to measure the density of the three solutions. To reinforce geometry, you can ask students to determine the density of their vials, using the formula for the volume of a cylinder (area of the base times height) and the mass. To challenge quick students, ask them to find the density of the vials two ways. (An alternative to using a formula is to determine the vial's volume by measuring how much water it displaces in a graduated cylinder.)

Answers to Student Questions

1. The bottom layer is cold salt water. The middle layer is cool freshwater. The upper layer is warm freshwater. All layers are clear.

2. Density of the water is responsible for the layering observed. Increasing the salt content of water increases its density. Increasing the temperature of the water decreases its density.

3. Answers will vary. Encourage students to conduct an honest evaluation of their hypothesis. Remind students that evaluation and revision are basic practices in all types of good investigation.

4. Allow students to experiment with this by leaving one apparatus alone for several days.

5. Answers will vary. Some strengths of this model include a layered system with little vertical mixing; layers based on temperature and salinity; layers based on relative densities; density differences based on differences in temperature and salinity. Some weaknesses of this model include ocean layers are not necessarily equal in thickness; except in some estuaries, there are no layers of freshwater in the ocean (except sporadic intense rainfall that creates freshwater lenses).

Assessment

- For a performance assessment, provide vials or test tubes with an intermediate number of BBs, and ask students where they predict it will float based on their group's data.

- If you grade students' answers to questions, allow students several days of observing one of their cylinders before answering question 4.

- For summative assessment, you can ask students to write a journal entry about what they know about the density of water—how it varies and what determines it, for instance. They could also make a concept map about the density of water. If they have done other Activities on density, including water density, ask them to include examples in their writing or concept map.

Activity 9 Summary

Students model the mixing of freshwater and salt water in an estuary, using a bread pan with layers of freshwater and salt water of different colors. Students first build the model and then jostle it to mix layers. They observe the model's response and reflect on how it compares to an estuary.

Activity	Subject and Content	Objective	Materials
Estuaries—Where the Rivers Meet the Sea	Mixing of estuaries	Investigate how water mixes in estuaries.	**Preparation:** Teacher will need: clear Pyrex glass loaf pans, "ocean water" (see Preparation section), "river water" (see Preparation section), pickling salt, food coloring (blue and yellow), graduated cylinder, pencil (or pen), masking tape, spoon, waste containers for used solutions, towels or rags for cleanup **Activity:** Each group of students will need: clear Pyrex glass loaf pan, 500 ml "ocean water"—the blue solution, 500 ml "river water"—the yellow solution, pencil (or pen), 20 cm masking tape, spoon, waste container for used solutions, blank paper, towels or rags for cleanup

Time	Key Concepts	Margin Features
Preparation: 25–30 minutes **Activity:** 50 minutes	II: Ocean structure and water movement III: Impact of human activities on the oceans	Safety Alert!, Fast Fact, What Can I Do?, Connections, Resources

Scientific Inquiry	Unifying Concepts and Processes	Personal/Social Perspectives
Modeling and observing	Modeling change within estuaries	Influence of human activity on estuaries

Estuaries— Where the Rivers Meet the Sea

Background

An estuary is a body of water partially enclosed by land that has a connection to a river or stream, and an opening to the ocean. Well-known examples in the United States are Chesapeake Bay, Puget Sound, Long Island Sound, and Galveston Bay. Many lagoons, saltwater marshes, and coastal wetlands are estuaries as well. They are places where freshwater coming from rivers and streams mixes with salty ocean water. The water circulation patterns within an estuary tend to trap nutrients—washed from the land and downstream through rivers and streams, creating a highly productive area for wildlife. Estuaries serve as nursery, spawning, and migration areas for many marine animals.

Many factors affect the mixing of salt water and freshwater in an estuary. Large differences in the amount of salt dissolved in ocean water as compared to river water slow down the mixing process. The range of the tides, the speed of the outgoing river/stream current, and weather conditions affect how quickly and thoroughly freshwater and salt water mix in an estuary. Scientists sampling estuarine waters sometimes find that the surface contains very little salt, but a wedge of very salty water lies on the bottom. In other cases, sampling indicates little variation in salinity from top to bottom.

In this Activity, you will examine a model of an estuary with distinct freshwater and saltwater layers. Pay particular attention to how easily the two layers mix.

Fast Fact

Estuaries are natural harbors and, therefore, were important sites for early human occupation. Estuaries are also natural breeding habitats for marine organisms that are sensitive to pollution associated with human activities. This conflict has led to the development of the United States Environmental Protection Agency's (EPA's) National Estuary Program, which was designed to conduct research geared to help sustain this important habitat.

Objective

Investigate how water mixes in estuaries.

Topic: deltas and estuaries
Go to: *www.scilinks.org*
Code: PESO 007

Activity 9

Materials

Each group of students will need

- clear Pyrex glass loaf pan
- 500 ml "Ocean Water"— the blue solution
- 500 ml "River Water"— the yellow solution
- pencil (or pen)
- 20 cm masking tape
- spoon
- waste container for used solutions
- blank paper
- towels or rags for cleanup

Time

50 minutes

Figure 9.1
Proper placement of the loaf pan on the pencil

Figure 9.2
Carefully pour the blue solution ("ocean water") into the loaf pan. Allow the water to become still before continuing.

SAFETY ALERT !

1. Be careful to quickly wipe up any spilled water on the floor—slip and fall hazard.

2. Wash hands with soap and water upon completing the lab.

Procedure

1. Using two pieces of masking tape, each about 10 cm long, tape the pencil to the lab table so that it cannot roll. (See **Figure 9.1**.)

2. Place the loaf pan on top of the pencil so that one end of the pan is higher than the other. (See **Figure 9.1**.)

3. Pour the blue solution ("ocean water") into the loaf pan. Allow the water to become still before continuing. (See **Figure 9.2**.)

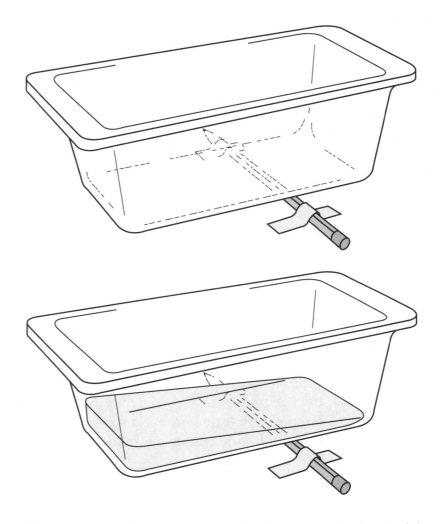

4. Hold the spoon so that it touches the inside of the raised end of the pan. Slowly and carefully pour the yellow solution ("river water") down the handle of the spoon so that it trickles into the upper end of the pan without splashing or disturbing the ocean water. (See **Figure 9.3**.) (Note: If the handles of the spoons are not suitable for use as described, then very gently pour the "river water" into the bowl of the spoon. It is important to cause little or no disturbance to the water already in the pan.)

Figure 9.3
Slowly and carefully pour the yellow solution ("river water") down the handle of the spoon.

5. Sketch and label—by color—the water layers you see in the pan. Describe how the "ocean water" and "river water" mixed, formed mixed layers, or stayed in separate layers as you poured in the "river water."

6. Using the pencil as a pivot point, rock the pan back and forth three times. Observe what happens. (See **Figure 9.4**.) This motion is more violent than what occurs in a real estuary, but simulates the mixing created by wind blowing over the surface of estuaries or by tidal currents flowing into and out of estuaries.

Figure 9.4
Rock the pan back and forth to simulate reversing tidal currents in an estuary.

7. Slowly stir the layers of water using the spoon. Observe what happens as you stir. What happens to the layers?

Activity 9

What Can I Do?

If you live near an estuary, volunteer with the United States Environmental Protection Agency's (EPA's) National Estuary Program to help clean up the pollution, much of which results from human activities in these densely populated coastal environments.

The EPA's Volunteer Estuary Monitoring Program allows everyone to get involved in helping to improve water quality and wildlife habitats.

Questions and Conclusions

1. What did you observe in step 6?

2. Based on your observations in this Activity, describe how the water in a real estuary might

   (a) form separate layers of fresh and salty water

   (b) become a "well-mixed estuary," having no separate layers

3. Based on your observations from this Activity and others you may have completed, what is responsible for the layering effect observed?

4. In what other areas of oceanography, or other areas of Earth science, does the layering of materials play an important role in the characteristics of our world?

5. What are the strengths and weaknesses of this experiment as a scientific model for investigating the structure of an estuary?

Estuaries— Where the Rivers Meet the Sea

What Is Happening?

This Activity demonstrates a phenomenon found in many estuaries: the overlaying of dense, salty ocean water by less salty water flowing into the estuary from rivers and streams. In some estuaries, the salty ocean water may form a well-defined, wedge-shaped layer on the bottom. This is commonly referred to as a "salt wedge" estuary. (See **Figure 9.5**.) The salt wedge remains distinct from the freshwater in this type of estuary.

The results that students obtain for this Activity will most often resemble the conditions found in a "partially mixed" estuary. In this type of estuary, the ocean and river layers are not sharply defined, but there is a definite difference in salinity between the top and bottom layers.

In some estuaries, the conditions are such that the entire water column has a uniform salt content from top to bottom. In "well-mixed" estuaries such as this, the salinity is less than that of the ocean, but greater than that of the rivers leading into the estuary.

Objective

Investigate how water mixes in estuaries.

Key Concepts

II: Ocean structure and water movement
III: Impact of human activities on the oceans

Materials (Preparation)

Teacher will need

- clear Pyrex glass loaf pans
- "ocean water" (see Preparation section)
- "river water" (see Preparation section)
- pickling salt
- food coloring (blue and yellow)
- graduated cylinder
- pencil (or pen)
- masking tape
- spoon
- waste containers for used solutions
- towels or rags for cleanup

Time (Preparation)

25–30 minutes

How Do We Know This?

Estuaries are sensitive to nutrient input from the adjacent landmass. How do we know this?

We know this because we can measure the amount of nutrients, such as nitrogen and phosphorus, in the water and bottom sediments. When high amounts of these nutrients are present, they create an increase in the amount of plankton, called a "plankton bloom." Although at first this may be considered a good thing, it really is not. This is for a variety of reasons, one of which is that the "plankton bloom" cuts off the amount of light reaching the estuary floor, killing many of the organisms living there.

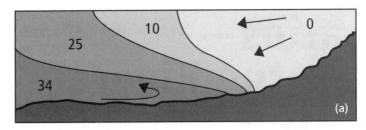

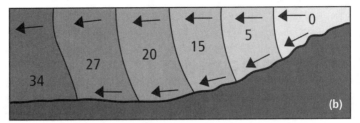

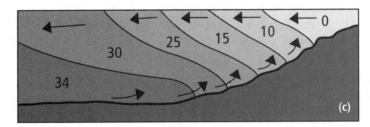

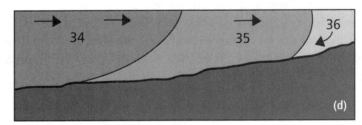

Figure 9.5
Schematic of the vertical cross section along the axes of four types of estuaries (from a physical oceanography perspective): (a) salt wedge, (b) well-mixed, (c) partially mixed, and (d) reverse. Salinity values in each type show the amount of mixing between freshwater (0 salinity) and seawater (35 salinity). Solid lines represent lines of equal salinity. The arrows indicate flow direction, and the thick black line is the bottom of the estuary.

Materials (Activity)

Each group of students will need

- clear Pyrex glass loaf pan
- 500 ml "Ocean Water"— the blue solution
- 500 ml "River Water"— the yellow solution
- pencil (or pen)
- 20 cm masking tape
- spoon
- waste container for used solutions
- blank paper
- towels or rags for cleanup

Time (Activity)

50 minutes

Preconceptions

Ask students, "What ideas do you have about what happens when river water flows into a bay or the ocean? Why do you think that? Give specific examples. What is the image you have in your mind about that?" Ask students to respond in their groups. Students might come up with the following:

- An estuary is the same as a bay in that it is simply a semienclosed body of ocean water with the same physical and biological characteristics as the open ocean.
- Estuarine waters are homogeneous with no vertical or lateral differences in salinity, temperature, or density.
- River or stream water mix instantly with seawater.
- Mixing never happens.
- Oceans are uniform throughout.
- Scientists have measured salinity and temperature throughout the ocean.

98

What Students Need to Understand

- The salt water from the ocean and the freshwater from rivers do not mix immediately in any estuary. There is always a tendency for the water to form layers, but in many estuaries, the currents, river flow, and tides mix the water very quickly.
- Estuaries are dynamic environments that can change rapidly as a result of storms, heavy rainfall inland in areas that drain into the estuary, or strong wind shifts. In addition to these unpredictable environmental factors, the water in estuaries is influenced by daily tidal changes.
- Estuaries are extremely important breeding areas for fish and other wildlife.

Time Management

This Activity can be completed in 50 minutes. Showing students the large demonstration column, having them experiment with their own column, and talking about this Activity may be done in 50 minutes. Preparing the demonstration will take approximately 25 to 30 minutes.

Preparation and Procedure

"River water" is simply tap water. Adding food coloring (3 to 5 drops) to the "river water" is not essential to performing this Activity, but the color contrast between yellow water of low salinity and blue water of high salinity makes the results easier for students to describe. Also, when "river" and "ocean" water mix, a layer of green water results.

The "ocean water" is prepared by adding 90 ml of dry pickling salt grains to 500 ml of tap water. Add 10 drops of blue food coloring to the solution. Stir until the salt is completely dissolved.

SAFETY ALERT

1. Be careful to quickly wipe up any spilled water on the floor—slip and fall hazard.

2. Wash hands with soap and water upon completing the lab.

Extended Learning

- For the same reasons that estuaries are so nutrient-rich, they are also areas that can become heavily polluted. They trap and store not only nutrients from land, but pesticides, oil, and chemicals that are in runoff and river and stream water. Preserving water quality in estuaries is a major concern of many environmental and sporting groups, such as the Nature Conservancy, Ducks Unlimited, and the Sierra Club. Investigate the work of these groups and the efforts of government and industry to deal with the pollution of estuaries if they are present in your state.
- Many states now have "critical area" laws that try to protect areas like estuaries from pollution and overdevelopment. Investigate whether your state or local area has any laws designed to protect estuaries.
- The nation protects estuaries through the Environmental Protection Agency's National Estuary Program (NEP). Local, regional, and federal agencies have

cooperated in developing a comprehensive plan for many sites, selected for their significance and threats. Ask students to study one of the sites and its management plan to learn about the features deemed worthy of protection, the threats to the site, and the process followed in developing the plan. Search for "EPA National Estuary" at *http://water.epa.gov/type/oceb/nep/index.cfm*.

- Another class of protected sites is the National Estuarine Research Reserve System (NERRS), a network created by the National Oceanic and Atmospheric Administration. The 27 NERRS sites are preserved for long-term scientific research projects. Students can search for the site closest to them, or one that intrigues them, and plan a trip to the site. What would they choose to look at and do there? What questions would they ask the researchers who work there? Search for NERRS at *http://www.nerrs.noaa.gov/*.

- If students live near the coast, encourage them to investigate a nearby estuary. They could participate in the Environmental Protection Agency's Volunteer Estuary Monitoring Program. This citizen science program's website includes instructions for specific protocols, many of which are suitable for middle and high school students. Search for "EPA Volunteer Estuary Monitoring" at *http://water.epa.gov/type/oceb/nep/monitor_index.cfm*.

- As a class, you can collect data on a local estuary using protocols designed for K–12 classes by the GLOBE Program. Students can collect data and observations on the water, atmosphere, soils, and flora by the techniques described on the website, and compare them over time and with other data collected by other students. Search for "GLOBE" at *www.globe.gov*.

- Students can compare their data to that collected automatically by buoys operated as part of the National Data Buoy Center's network. Data are posted in real-time on the website. Search for "NDBC" at *http://www.ndbc.noaa.gov*. Another source of real-time data is nowCOAST, from the National Oceanic and Atmospheric Administration. This site can introduce students to Geographic Information Systems (GIS) as they examine estuarine and meteorological data in layers. Search for "nowCOAST" at *http://nowcoast.noaa.gov*.

Connections

Estuaries are critical to the health of our coastal environment. The National Estuary Program, part of the EPA, was established by congress in 1987 to improve the water quality in estuaries of national concern. Search for the National Estuary Program at *http://water.epa.gov/type/oceb/nep/index.cfm*.

Interdisciplinary Study

- Have students study and research painters who have painted seascapes, marshes, and estuaries. Paintings of waterfowl, many in estuarine areas, might also be researched. Ask students to compare and contrast the styles, methods, and composition of various painters.

- Have students read a portion of Homer's *Odyssey*, Book 5. Search online for "Odyssey 'Struggling, he grasped the rock.'" Begin with line 424, "While he thus doubted in his mind and heart." Ask students to respond in small groups to the piece. What are Odysseus's senses detecting? (What is he hearing, tasting, feeling, etc.?) How is he feeling emotionally? What words or phrases give clues about how he is feeling? What images does the passage evoke? How does this passage relate to estuaries?

- Students can also study organisms that are adapted to survive and breed in estuaries, where the salinity, temperature, and density of water vary. What are some physical characteristics that allow these organisms to survive, and what behaviors do they exhibit?

Differentiated Learning

For avid readers, suggest they read William W. Warner's book, *Beautiful Swimmers: Watermen, Crabs and the Chesapeake Bay*. Or, for the benefit of reluctant readers, consider reading this book or portions of it aloud. The book describes how the lives of watermen and blue crabs have historically intertwined in the Chesapeake.

Answers to Student Questions

1. Students might report that they saw water sloshing, colors bleeding into each other slowly, or colors blending to make green.

2. (a) A steady flow of freshwater into a calm, protected estuary will tend to produce well-defined layers of water of varying salt concentration.

   (b) Estuaries influenced by strong tides, high winds, and variable freshwater flow will tend to become well mixed.

3. Answers may vary, but students should be able to recognize that salt water is denser than freshwater and that this density difference is responsible for the layering effect.

4. Answers will vary. Some possibilities include layers within the open ocean, Earth itself, the atmosphere, rocks, and so on.

5. Answers will vary. Some strengths of this experiment as a scientific model include a layered system with some mixing; water containing differing amounts of dissolved salts do not readily mix. Some weaknesses include no analogue in nature for the rocking of the loaf pan; an oversimplified system—real estuaries are extremely complex.

Resources

http://water.epa.gov/type/oceb/ nep/index.cfm

www.nerrs.noaa.gov

http://water.epa.gov/type/oceb/ nep/monitor_index.cfm

www.globe.gov

www.ndbc.noaa.gov

http://nowcoast.noaa.gov

Assessment

- For formative assessment, circulate as students do the Activity and ask them open-ended questions such as, "What do you see happening? Why do you think it's happening? Are there any variations that you'd like to try? Why? What do you expect to happen?"
- For summative assessment, besides grading answers to questions, you can ask students to write about how their understanding of estuaries has evolved by doing this Activity. Ask them to describe or draw the setting of an estuary and to describe in particular what happens when freshwater meets salt water.

Activity 10 Planner

Activity 10 Summary

Students build a model of the Atlantic basin with clay in a baking pan. They create currents by blowing through a straw to explore how wind and the shape of the basin affect currents. Questions guide students to relate their model to currents in the Atlantic Ocean.

Activity	Subject and Content	Objective	Materials
Current Events in the Ocean	Surface ocean currents	Model how landforms and wind affect ocean surface currents.	Each group of three to four students will need: indirectly vented chemical splash goggles and aprons; baking pan, 30 cm × 45 cm × 3 cm (12 in. × 18 in. × 1.5 in.) deep, painted black inside; white chalk; modeling clay; colored pencils; a plastic drinking straw with a flexible elbow for each student; black permanent marker (no or low VOC); 400 ml rheoscopic fluid; towels or rags for cleanup

Time	Vocabulary	Key Concept	Margin Features
50 minutes	Ocean currents, Gulf Stream, California Current, Rheoscopic fluid	II: Ocean structure and water movement	Safety Alert!, Fast Fact, What Can I Do?, Connections, Resources

Scientific Inquiry	Unifying Concepts and Processes
Modeling and observing	Change due to atmosphere/ocean interactions

Current Events in the Ocean

Background

Sailors have known for centuries that **ocean currents** can speed up or slow down a ship. In modern times, scientists have discovered that ocean currents have major effects on climate and weather patterns, and on the ecology of the ocean and nearby landmasses. One type of current is called a surface current. The **Gulf Stream**, which flows along the east coast of the United States, and the **California Current**, which flows along the west coast of the United States, are both surface currents. Surface currents, as you might have guessed, flow across the surface of the ocean almost like a river flows across dry land. However, a surface current does not have solid banks like a river to direct its flow. As a result, the direction of a surface current may change when the wind blowing across it shifts, when it encounters warmer or colder water, or when it nears land.

Vocabulary

Ocean currents: Regular movement of large water masses in the ocean along certain defined paths. In this Activity, we are dealing with surface currents, but there are other types of ocean currents.

Gulf Stream: A specific ocean current that flows south to north along the east coast of the United States. It carries warmer water from lower latitudes into and across the Atlantic Ocean.

California Current: A specific ocean current that flows north to south along the west coast of the United States. It carries cooler water from the northern areas of the Pacific Ocean south toward the equator.

Rheoscopic fluid: Fluid capable of effectively helping in the visualization of dynamic currents using suspended particles in a liquid matrix.

Objective

Model how landforms and wind affect ocean surface currents.

Topic: ocean currents
Go to: *www.scilinks.org*
Code: PESO 008

Activity 10

Materials

Each group of three to four students will need

- indirectly vented chemical splash goggles and aprons
- baking pan, 30 cm × 45 cm × 3 cm (12 in. × 18 in. × 1.5 in.) deep, painted black inside
- white chalk
- modeling clay
- colored pencils
- a plastic drinking straw with a flexible elbow for each student
- black permanent marker (no or low VOC)
- 400 ml **rheoscopic fluid**
- towels or rags for cleanup

Time

50 minutes

Figure 10.1
Firmly press the clay in the pan to create a boundary system to contain the "ocean."

Fast Fact

The consistently fastest surface ocean currents are the Gulf Stream off the east coast of the United States, the Kuroshio Current off Japan, and the Agulhas Current in the Indian Ocean off South Africa. They have maximum surface velocities of more than 2 m per sec., which is as fast as many major rivers.

Procedure

1. Put on indirectly vented chemical splash goggles and aprons.

2. Bend the straw at the elbow. Write your name on the short end of the straw, using the black permanent marker. This will allow you to tell which straw is yours; it will also remind you which end of the straw to blow into. Do not put the end of the straw with your name written on it into your mouth.

3. Using **BLM 10.1** (which shows the positions of the continents), draw an outline of the eastern coasts of North and South America and the western coasts of Africa and Europe inside the baking pan with the white chalk. Also draw an arrow indicating north inside the pan, but not in the area representing the Atlantic Ocean.

4. Following the chalk pattern, place ridges of modeling clay along the bottom of the pan to create a boundary system to contain the "ocean." Press the clay firmly to the pan and smooth the gaps between the clay and the pan as shown in **Figure 10.1**. It is important to create a watertight seal to prevent "oceanic" leaks.

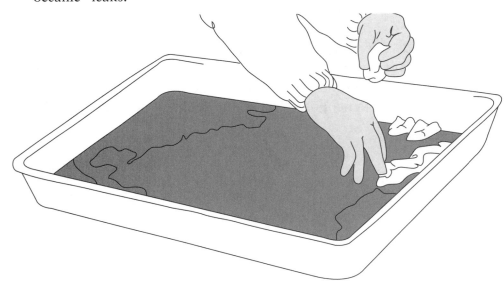

5. Fill the ocean area (center) of your model with rheoscopic fluid. Wait for the swirling patterns in the solution to settle.

   Try not to touch the tray as you do the following steps:

6. For centuries, sailors have used the trade winds to cross the Atlantic from Europe and Africa to Central and South America. Simulate these trade winds as follows:

   Hold your straw so that the short end (with your name on it) is parallel to the ocean surface. Point your straw from the bulge of Africa westward toward the coast of South America. Gently blow through the straw and observe the patterns of ocean currents that the wind produces. (See **Figure 10.2**.)

Figure 10.2
Gently blow through
the straw and observe
the patterns of water
movement (ocean currents).

7. Using one colored pencil, sketch the swirling patterns produced by the wind on Map 1 of the data sheet (**BLM 10.1**). With a different colored pencil, use a single arrow to indicate the direction your straw was pointing when you blew. Look at the current that forms in your model. Compare this to the Gulf Stream, which runs along the east coast of the United States.

8. Alter the direction of the straw, pointing it more northerly or southerly, and gently blow through the straw. Observe how the current patterns change. On Map 2 of the data sheet (**BLM 10.1**), sketch the current patterns observed and a single arrow to indicate wind direction. Pay particular attention to what happens as the fluid flows around landmasses like Florida.

9. Repeat step 7 at least three times (more if time allows) to develop a sense of the relationship between wind direction, landmass configuration, and current patterns. You may wish to put small islands into your ocean and observe their effects on current flow.

10. Repeat step 5, noticing the region in the middle of the Atlantic Ocean. This region is known as the "Sargasso Sea." Discuss with your group the characteristics of currents in that region.

Questions and Conclusions

1. Describe the relationship between wind and ocean currents. What variables, such as wind direction, wind speed, land formations, and so on, seem to be related to ocean current patterns?

SAFETY ALERT

1. Indirectly vented chemical splash goggles and aprons are required.

2. Wash hands with soap and water upon completing the lab.

3. Use only GFI-protected circuits when working with water or other liquid near electricity.

4. Never share with other students equipment that you put your mouth on (e.g., straws).

5. Pertinent safety procedures and required precautions outlined in MSDSs for hazardous materials will be reviewed with you by your teacher.

Activity 10

What Can I Do?

If you live near the ocean or have a chance to visit it, you can look for the effects of ocean currents when you are on the beach. Look for things that have washed up from ships. Classic examples are messages in bottles, running shoes, and fishing floats. You can also track ocean currents by watching oceanographers' drifter buoys online. Pick three or four buoys and set up a fantasy relay team to race against a friend's team for the longest distance traveled in a month. You will find the buoys by searching for "GDP Drifter" at *http://www.aoml.noaa.gov/phod/dac/dacdata.php*. You can then use Google Maps or Google Earth to find the buoys.

2. In the late 1700s, Benjamin Franklin described a warm surface current of water running across the Atlantic Ocean from the east coast of North America toward England. We now know this to be the Gulf Stream. Based on the observations you have made with this model and from your knowledge of how water masses containing differing amounts of salt form layers, explain why the warm water of the Gulf Stream stays as a distinct current as it moves northward through the colder, saltier Atlantic Ocean waters.

3. The latitude of Edinburgh in Scotland is approximately that of Moscow in Russia, yet their climates are quite different—Edinburgh experiences relatively mild winters compared with the winters in Moscow. How can this be related to the Gulf Stream? Explain your answer.

4. This Activity uses a simple model of the Atlantic Ocean and the ocean currents that flow in it. Although scientists use models to help them answer complicated questions such as "How do currents form?", the models are limited. Think about how your model is like a real ocean and how it is different. What are some strengths of your model? What are some weaknesses?

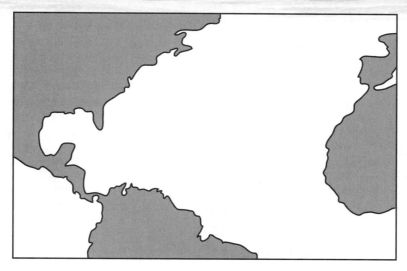

Map 1

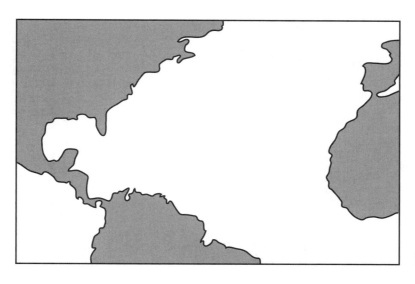

Map 2

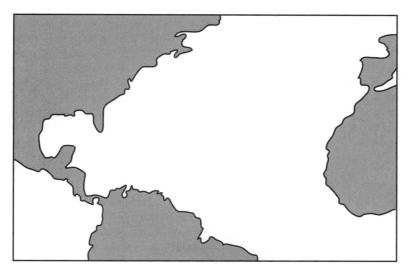

Map 3

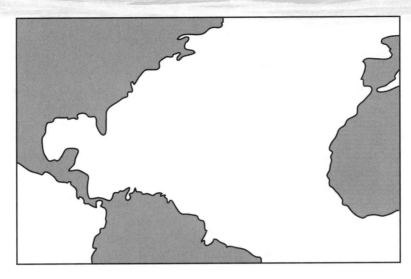

Map 4

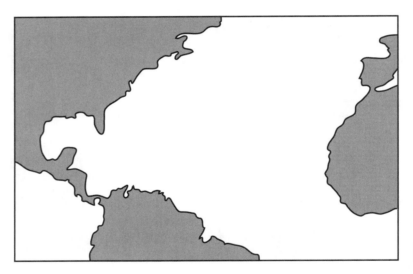

Map 5

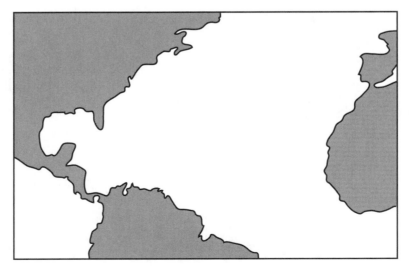

Map 6

Current Events in the Ocean

What Is Happening?

Currents, large-scale movements of water, occur throughout the ocean—in the surface layers as well as in the subsurface layers. Surface currents are driven primarily by wind. A map showing prevailing wind direction will correlate remarkably well with a map showing surface currents in the major ocean basins. (See **Figures 10.3** and **10.4**.)

This Activity allows students to simulate the flow of the Gulf Stream in the Atlantic Ocean and experiment with how land formations and variations in wind direction can affect ocean currents. The Gulf Stream actually flows from south Florida northeastward toward Cape Hatteras, North Carolina. It then curves eastward and flows across the North Atlantic toward Ireland and Great Britain. The friction from permanent wind systems—the trade winds in the south and the prevailing westerlies in the north—appears to be the most important factor in determining the Gulf Stream's motion. It is important, however, not to mislead students into thinking that major current systems are simply the result of wind blowing water in a certain direction. It is in reality much more complex, and there are many factors that determine current formation in the ocean. For instance, we have looked at some of the effects of density in previous Activities and we will study the effect of tides in Activities 14 to 16.

How Do We Know This?

How do drifter buoys work and how do they send back their signals?

Drifters are one of the oldest methods used to study how the ocean moves. Surface drifters float near the surface of the ocean so the currents transport them, but without having too much area over the surface; otherwise, wind would push them and they would be measuring the wind rather than the ocean. Other drifters travel in the interior of the ocean, some at a depth of 1,000 m. These drifters come to the surface every 10 days, measuring as they move up to provide information on the vertical structure of the ocean. In most cases, drifters transmit their information to satellites, which send the data back to the scientists. Some drifters that travel in the interior of the ocean send acoustic pulses that are received by hydrophones (i.e., microphones with very long ranges) near the coast.

Objective
Model how landforms and wind affect ocean surface currents.

Key Concept
II: Ocean structure and water movement

Materials
Each group of three to four students will need

- indirectly vented chemical splash goggles and aprons
- baking pan, 30 cm × 45 cm × 3 cm (12 in. × 18 in. × 1.5 in.) deep, painted black inside
- white chalk
- modeling clay
- colored pencils
- a plastic drinking straw with a flexible elbow for each student
- black permanent marker (no or low VOCs)
- 400 ml rheoscopic fluid
- towels or rags for cleanup

Time
50 minutes

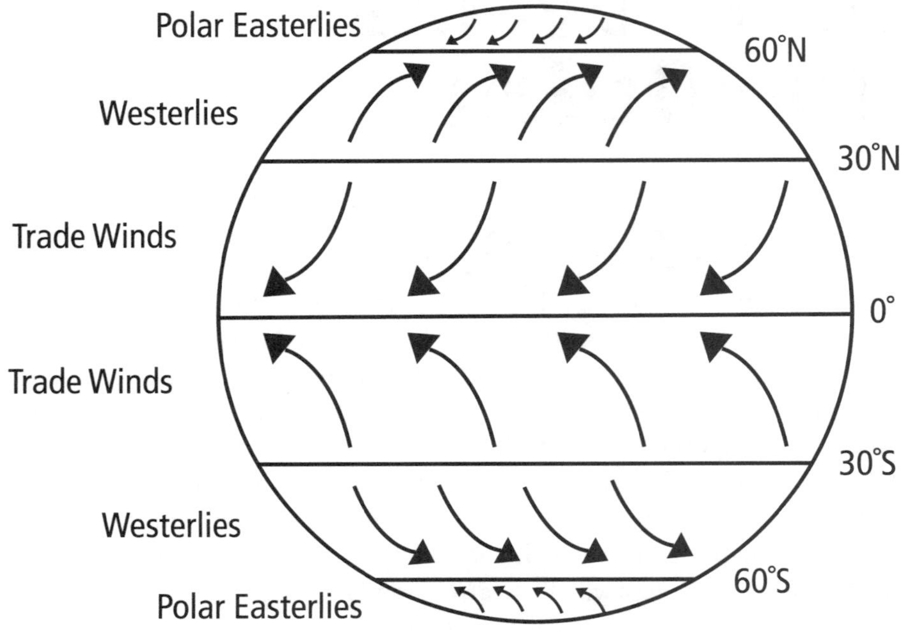

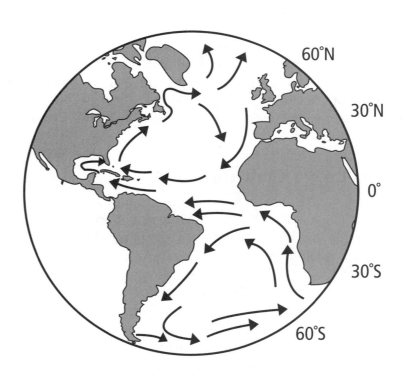

Figure 10.3
Global wind
circulation patterns

Figure 10.4
Surface ocean currents
in the Atlantic Ocean

Unlike rivers on land, ocean currents may shift their course when influenced by outside factors. These shifts in turn can lead to a variety of changes around the world. In the last 30 years, scientists have been learning more about the cold Pacific Humbolt current—which flows northward along the west coast of South America—and a shift in it that is associated with the phenomenon known as "El Niño." This is produced by a weakening or change in the direction of the trade winds along the equatorial Pacific that allows warm water to "slosh" eastward. El Niño has greatly disrupted the South American fishing industry and caused much loss of bird life in nesting sites on small Pacific islands. In the Atlantic Ocean, a shift in the path of the Gulf Stream carried fish that are usually found only in the Caribbean Sea thousands of miles north to the coast of North Carolina.

The reflective properties of the particles suspended in the rheoscopic fluid allow students to see subtle aspects of the currents in a model of the Atlantic Ocean and its adjacent continents. By blowing gently across the model's surface, students generate tiny currents that simulate the huge ocean currents moving millions of cubic meters of water across an ocean.

Preconceptions

Ask students if they have ever heard of unusual objects washing up on shore and, if so, could they tell where the object came from. There are the classic stories of messages in bottles, but on more than one occasion, containers from ships have washed overboard, spilling running shoes (1990 in the north Pacific) and rubber ducks (1992 in the western Pacific) that have literally traveled the world. Ask students what they have learned about ocean currents such as the Gulf Stream. Students may express some of the following beliefs:

- Currents are driven by wind alone.
- Currents are only on the ocean's surface.
- Currents are constant, staying in the same place and with the same velocity.

What Students Need to Understand

- Surface currents in the ocean are driven primarily by wind.
- While surface currents, like the Gulf Stream, are primarily affected by wind direction, other factors, such as land formations, influence surface currents.
- The prediction of when and how ocean currents change course is as complex as weather prediction. Occasional short-term current shifts occur all over the ocean and can lead to major changes in weather patterns, fish harvests, and so on.

Time Management

This Activity can be completed in 50 minutes. Be sure to allow time for students to clean up. If they get engaged in the model, be prepared for them to need time to answer questions and discuss their results the following day.

Preparation and Procedure

Spray paint the interior of the trays used for this experiment with a flat black, enamel, rustproof paint. This type of paint is available at most hardware and paint stores. Allow at least one hour for the paint to dry. Other methods of painting or types of paint may be used, but make sure the paint dries with a flat or matte finish, not glossy. A black cafeteria tray may also work, although take care to make sure the clay is built up high enough to prevent the fluid from spilling.

Rheoscopic fluid is ideal for demonstrating a variety of currents, including convection currents, aerodynamic flow, and much more. It can be reused year after year. (See the Resources list for information on how to obtain rheoscopic fluid.)

While the rheoscopic fluid is nontoxic, you should require students to wear indirectly vented chemical splash goggles while performing this experiment. As with most foreign materials, should eye contact occur, flush with plenty of water.

Prior to doing the Activity, review with students pertinent safety procedures and required precautions for the rheoscopic fluid according to the MSDSs.

Remind students not to share straws.

Moisten the clay before cleaning up. Clay dust contains silica and can be a respiratory health hazard.

Alternative Preparation

As an alternative to rheoscopic fluid, you can substitute water and food coloring.

If you are going to use water and food coloring, you should not paint the aluminum pans black. Paint them flat white instead and the food coloring will show up much better. The procedures will need to be changed as follows: After creating the outline of the continents with the clay, pour water into the area representing the Atlantic Ocean. Put a drop of food coloring in the water where you will begin blowing. Be sure to remind students to watch the direction in which the food coloring moves. The food coloring will dissipate quickly, so be careful to have students watch closely. Another alternative is to use water and tiny Styrofoam packaging pieces.

As another alternative, or for demonstration purposes, you may wish to use a hair dryer to create the "wind". (See **Figure 10.5**.) Select a dryer that allows you to use a low or no heat setting and a low fan speed. *Use extreme caution if you choose to use an electrical appliance near water. Plug the hair dryer into only a GFI-protected circuit.* Use a stand to hold the hair dryer in place, at least 15 cm from the fluid. Students should use appliances only with close adult supervision. (See the Safety Alert on this page.)

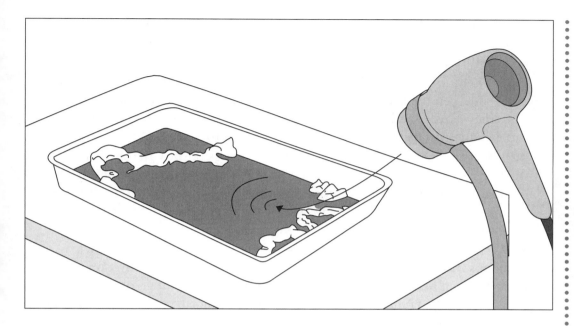

Figure 10.5
Instead of having students blowing through a straw to create the "wind," use a hair dryer set at low speed and heat.

Extended Learning

• Have students make models of other continents and adjacent portions of the ocean. See if you can generate currents in the rheoscopic fluid that simulate known ocean currents. Students also can make islands or jetties along a simulated coastline to study how these can affect longshore current flow.

• Ask students to explore drifter buoys—buoys that oceanographers deliberately drop overboard to measure currents, temperature, and salinity. Some of these are designed to stay at the surface, while others sink to a specific depth to collect data. Woods Hole Oceanographic Institute runs several programs with drifter buoys. Search for "whoi drifter buoys" at *www.whoi.edu/page. do?pid=10320.* For buoys that float below the surface, search for "whoi ALACE drifter" at *www.whoi.edu/page.do?cid=9706&pid=11696&tid=282* or "whoi RAFOS drifter" at *www.whoi.edu/instruments/viewInstrument.do?id=1061.* See What Can I Do? for data from buoys.

Interdisciplinary Study

• Encourage students to conduct library research on a number of topics related to the concepts presented here. Some possibilities include the effects of El Niño in the Pacific Ocean on weather patterns around the world; the ecosystem in the Sargasso Sea and its relationship to ocean surface currents; and the ecological and commercial significance of the Gulf Stream. Students can investigate explorers' routes and how and why they took those routes. Did currents affect their choices? Look at Columbus's voyage to India that resulted in his landing in

Connections

Other Earth scientists also investigate currents. Geologists study the motion of lava, magma, and even solid rock in the mantle that flows slowly over time. Meteorologists follow the movement of jet streams to help predict weather. Ask students to investigate one of these. They can, for instance, do "Solid or Liquid? Rock Behavior Within Earth" from *Project Earth Science: Geology*. They could study the jet stream by following it on television weather broadcasts or by searching for "jet stream forecast," which results in websites with maps that allow students to compare predictions made by commercial weather forecasters.

the islands of the Caribbean. What would have happened if ocean currents flowed in the opposite direction? How did he return to Europe and what current or currents aided his return voyage?

- Many cultures and peoples, especially cultures that have traditionally depended on the ocean for transport and food—like peoples of the South Pacific—have used knowledge of currents for centuries. Explore some of these cultures' methods of transport and trade routes.

- Investigate paintings of the sea. Do the paintings portray the sea as calm and peaceful, or stormy and threatening? When might the sea appear tranquil and unthreatening, or stormy and treacherous? What forces in nature affect the state of the sea? What are possible dangers of a sea that is stormy? Who is most affected by the "moods" of the sea?

- One painter you may want students to investigate is Winslow Homer, a famous American artist of the 19th century. Homer portrayed dramatic moments in human interaction with the mighty forces of nature found in the ocean. His favorite motifs included small sailing craft in which the crew was pitted in a struggle against the elements of the sea.

Differentiated Learning

This Activity uses models to illustrate the effect of wind on currents, and therefore should be successful with a broad range of students. You can ask students who enjoy mathematical challenges to determine the scale of their models. With highly mathematically gifted students, you can introduce vectors, using current velocity as examples. (Vectors are typically introduced in Algebra 2.)

Answers to Student Questions

1. Wind flow determines current flow in open areas of the ocean and can be instrumental in maintaining flow. Increasing wind speed can increase current speed. Continents or islands can alter the direction of current flow.

2. Because the Gulf Stream is composed of warm, less dense water, it does not mix readily with the cold, dense water flowing down the east coast from the North Atlantic.

3. The Gulf Stream crosses the Atlantic Ocean as a distinct mass of warm water and moderates the climate of the British Isles.

4. Answers will vary. Strengths might include the following: my model makes a complex system simple, so I can observe what is going on; it lets me change wind patterns to see how different ones affect ocean currents; it creates a good model of how the Gulf Stream flows. Weaknesses might

include the following: my model is much smaller than the real thing; factors other than wind cannot be studied; it uses a fluid, not real salt water; it is only one depth, while the real ocean varies in depth; the bottom of the pan might influence currents at the surface.

Assessment

- Monitor students as they do this Activity and encourage them to modify the model. They should articulate what they would like to do, why, what they think they will see, and then what they did see happen. Ask students to share their results with the class.
- For summative assessment, you can ask students to summarize in a few sentences what they have learned about surface currents. Ask them what they learned that surprised them, and why. Or, you could grade students' answers to questions.

Resources

www.aoml.noaa.gov/phod/dac/dacdata.php

www.whoi.edu/page do?pid=10320

www.whoi.edu/page.do?cid=9706&pid=11696 &tid=282

www.whoi.edu/instruments/viewInstrument.do?id=1061

Activity 11 Planner

Activity 11 Summary

Students become a wave in this quick Activity. A line of students stands side by side. Another student starts a wave by pushing or pulling forward the first student in line, sending energy down the line.

Activity	Subject and Content	Objective	Materials
Body Waves	Wave motion and energy	Investigate the energy of a wave and the motion of the medium through which a wave travels.	None

Time	Vocabulary	Key Concept	Margin Features
30 minutes	Oscillate	II: Ocean structure and water movement	Safety Alert!, Fast Fact, What Can I Do?, Connections, Resources

Scientific Inquiry	Unifying Concepts and Processes
Modeling and Observing	Energy flow via waves

Body Waves

Background

Waves are among the most common phenomena in nature. Waves breaking in the ocean, sound, light, microwaves, radio, or the motion of a guitar cord or a drum are all forms of waves. Such diversity makes finding common properties a challenge but, in general, waves can be explained as a disturbance that moves from one place to another place. In most cases, the disturbance propagates through a medium (even though electromagnetic waves such as light, radio, or microwaves can travel through empty space), but the medium itself does not move in the same way as the wave.

In the case of ocean waves, the medium is water. If you stand on a pier at the ocean and watch the waves, it is clear that they move toward the shore. This can even be seen in a small pond when a rock is tossed in the middle. While ocean waves travel toward the shore, the water they travel through remains relatively still, moving mostly up and down. People who have floated in the ocean out beyond the breakers know this from firsthand experience.

There are two types of waves depending on the relation between the direction of motion of the medium and the direction of propagation of the wave: transverse and longitudinal. Transverse waves, as shown in **Figure 11.1**, are the most familiar. Ocean waves are generally classified as transverse waves. In a transverse wave, the medium—water, in the case of an ocean wave—**oscillates** perpendicular to the wave direction. In an ocean wave, the *wave* moves toward the shore and the *water* moves up and down, perpendicular to the motion of the wave. Light also travels in transverse waves.

Vocabulary

Oscillate: To move or travel back and forth between two points.

Fast Fact

The highest ocean wave ever measured from storms was 28 m (91 ft.) in the Gulf of Mexico during Hurricane Ivan in 2005. That wasn't even the worst of the storm; the measuring system was shut off during the highest waves, which models estimated to be 40 m (130 ft.) high. The highest wave outside of a storm not caused by a tsunami was a "rogue wave" measured by *USS Ramapo* in 1933 that was 34 m (112 ft.). The highest wave caused by a tsunami occurred in Lituya Bay, Alaska, in 1958, and measured 530 m (1,740 ft.).

SCI**LINKS**
THE WORLD'S A CLICK AWAY

Topic: ocean waves
Go to: *www.scilinks.org*
Code: PESO 009

Objective

Investigate the energy of a wave and the motion of the medium through which a wave travels.

Activity 11

Materials

None

Time

30 minutes

The other type of wave is longitudinal (or compression). Longitudinal waves can be represented by the compression and expansion of a spring, as shown in **Figure 11.2**. Sound travels in longitudinal waves. You may have noticed that very loud sounds can actually hurt your ears. This is because the air in which the sound travels pushes against the eardrum, causing pain. In a longitudinal wave, the medium oscillates parallel to the wave direction. (See **Figure 11.2**.)

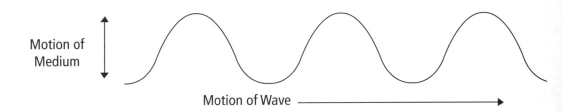

Motion of Medium

Motion of Wave

Figure 11.1
Transverse wave

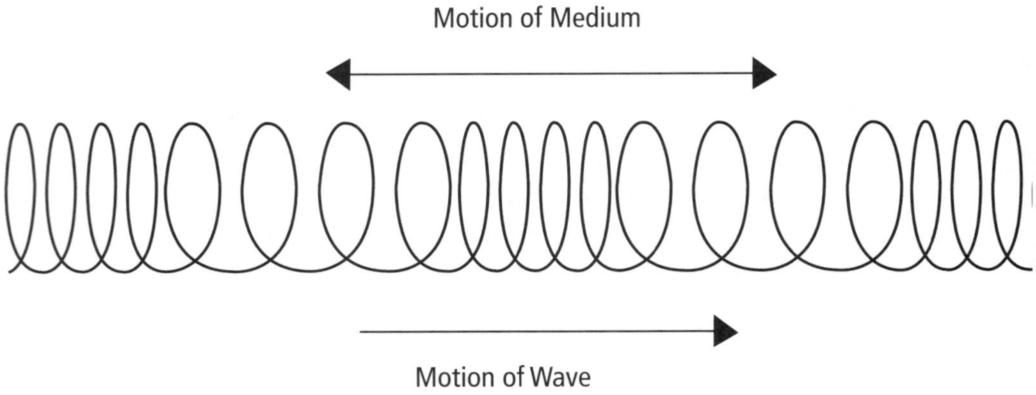

Motion of Medium

Motion of Wave

Figure 11.2
Longitudinal wave

Procedure
Part 1

1. Form groups of 7 to 12 people. Decide among you who will be "the force" (just one of the members of the group), and who will be part of "the wave" (the rest of the group).

2. The members of "the wave" should stand shoulder to shoulder in a straight line, all facing the same direction, preferably toward the rest of the class. You should be standing "at ease," not bracing yourselves or pushing against each other, with each shoulder resting against the shoulder of the student on either side of you. (See **Figure 11.3**.)

Figure 11.3
Students standing
shoulder to shoulder

3. "The force" should stand at one end of the line and give the end student a gentle shove toward the other end of the line. Repeat at least two more times and make more observations.

Part 2

4. The members of "the wave" should now stand side by side all facing the same way and interlocking their arms, as illustrated in **Figure 11.4**.

Figure 11.4
Students standing with
interlocked arms

5. "The force" should stand at one end of the line and pull the end student forward in the direction he or she is facing, and then back. Do this several times until every student in the line is moving. Describe the motion of students in the line. In what direction did they move? What was the direction of the wave? What made them move? Where did the energy come from? Where did the energy go? What kind of wave is this? Repeat at least once and make more observations.

Activity 11

What Can I Do?

If you live near water, you can estimate or measure several features of waves and see how they change as weather and tides change. Wave height (also called "amplitude") is the distance from crest to trough; you can estimate this against a vertical structure like a piling or pier. The wavelength is the distance between two crests; you can estimate this against a long horizontal structure like a jetty, wharf, or pier. The wave period is the time that it takes for two wave crests to pass the same point; you can time this with a watch or phone as waves pass by a piling, pier, buoy, or boat. If you live inland, you can monitor waves at your favorite beaches on the National Data Buoy Center website. Search for "NDBC" at *www.ndbc.noaa.gov/*.

Questions and Conclusions

Part 1

1. What happened to the students all the way down the line?
2. Where did the energy to make them move come from?
3. How far did the energy travel?
4. Did each student move as far as the energy traveled?
5. What kind of wave was this?
6. Other observations:

Part 2

7. Describe the motion of the students in the line. In what direction did they move?
8. What was the direction of the wave?
9. What made them move? Where did the energy come from?
10. Where did the energy go?
11. What kind of wave is this?
12. Other observations:

120

National Science Teachers Association

Body Waves

What Is Happening?

Students who have had the opportunity to go to the ocean or a large lake have an advantage in understanding some of the concepts related to waves. Two such concepts are (1) the medium in which a wave travels (water, in the case of ocean waves) does not move in the same way as the wave itself, and (2) waves transfer, or carry, energy.

Both transverse and longitudinal waves can carry and transfer tremendous amounts of energy. Ocean waves provide an excellent example of this. They are capable of knocking people down who are standing in the surf, capsizing huge ships at sea, and transferring massive quantities of sand.

In some locations around the world, wave energy is actually converted into electricity. This Activity allows students to simulate longitudinal and transverse waves, and feel the energy these waves transmit.

Objective
Investigate the energy of a wave and the motion of the medium through which a wave travels.

Key Concept
II: Ocean structure and water movement

Materials
None

Time
30 minutes

How Do We Know This?

How do we know that the medium through which waves travel does not have net movement?

An easy way to show that waves do not carry matter when they move is to do an experiment with a wave tank filled with rheoscopic fluid (for a definition of this kind of fluid, see Activity 10). As the wave propagates, the particles near the surface move in a full circle, while the bottom particles just move back and forth. If you were to repeat the experiment in sufficiently deep water, then as the wave passes, the particles move in full circles near the surface. The size of the circle will decrease as we go down to a certain depth at which the particles do not move at all.

See *www.classzone.com/books/earth_science/terc/content/ visualizations/es1604/es1604page01.cfm?chapter_no=visualization.*

Preconceptions

You can ask students to tell you what they have learned in school about waves and how that compares to their personal experience with water waves. You can collect and display their responses for the class to see. Another option is to ask them what they think they know about waves and if there is anything they would like to learn. Here are some possible preconceptions students might have:

- As a wave moves through matter, the matter itself moves long distances.
- Wind causes all water waves.
- Every seventh wave is a big one.
- Waves are largest near shore.
- Waves only break near shore.

This option uses the first two parts of a research-based strategy called KWL: What do you **K**now? **W**hat would you like to know? And, at the end of the instruction, what did you **L**earn? (See **Figure 11.5**.)

Topic: _____		
What Do You **K**now?	**W**hat Would You Like to Know?	What Did You **L**earn?

Figure 11.5
Numerous KWL graphic organizers such as this are available online by searching for "KWL chart."

What Students Need to Understand

- While a wave moves in one direction, there is little movement in the medium through which it moves.
- Waves require energy to begin.
- Waves can transfer energy over long distances.

Time Management

This demonstration takes only a few minutes to do, but allot at least 30 minutes so that students have time to write their preconceptions, do the Activity, and respond to questions about the Activity.

Preparation and Procedure

The only preparation required is to make sure there is enough open space in the classroom to do the demonstration safely. Remove any furniture, equipment, or other items that may be in the path of students doing this Activity. Also watch for trip and fall hazards. If the classroom is too crowded, the demonstration could be done outdoors, in a gym or multipurpose room, or in the hall.

> **SAFETY ALERT** !
>
> To prevent injuries, make certain there is no furniture, equipment, or other items in the area of this Activity.

Extended Learning

Challenge students to think of other ways to demonstrate that the water in a wave does not move in the direction of a wave. They can build a wave tank to show this. (See Activity 12.) Students may suggest that the popular "stadium wave" is a good model of an ocean wave but, in fact, it is not. Ask them why. The medium (people) in which a stadium wave travels is not continuous, whereas water is. In a stadium wave, each person has to provide his or her own energy to stand up and sit down. In an ocean wave, the water particles transfer energy from one particle to another. Refer to the Activity you just did and the fact that energy transferred from student to student—each student did not have to use his or her own energy to continue the wave.

> **Connections**
>
> Earthquakes propagate through the solid earth via waves of energy. Seismologists quickly determine where an earthquake originated—the focus or hypocenter—by comparing the time elapsed for a sequence of different kinds of waves to reach different spots around the globe. Ask students to investigate how seismologists do this. A good starting point with a real example is the U.S. Geological Survey's Earthquake Hazards Program FAQ Measuring Earthquakes. Search for "USGS finding earthquakes" at *http://earthquake.usgs.gov/learn/faq*.

Interdisciplinary Study

- Encourage students to study the impact of wave energy. Look at how coastlines erode due to wave action. Beach erosion is a topic around which there is much debate. Should beaches be replenished or should we let nature take its course? Are there ways to protect beaches from waves and erosion? If you live in a coastal state, have students learn what your state or community's policy is about coastal development, beach replenishment, or armoring of beaches with hard surfaces. If nearby beaches have been replenished, have students find out the cost and discuss whether the economic, social, and recreational benefits warrant the expense. Students could debate this, write a letter to the editor, or write a letter to their city councilmember or legislator.

- Ludovici Einandi's composition, "Le Onde," or "The Waves," is a 1996 short piece of classical piano music available in numerous recordings on YouTube. You can play this four-minute piece and ask students to describe the images and feelings the piece evokes for them.

- Students can explore frequency and amplitude of waves with a simulation about sound waves from the University of Colorado's Physics Education Technology (PhET) website. Search for "PhET," then search the site for "sound" at *http://phet.colorado.edu*.

- Waves can be a writing prompt for a journal exercise. Teachers Domain carries a half-minute visual recording of waves with a built-in writing prompt about the interactions between oceans and shores. Search for "wave action" in Teachers' Domain at *www.teachersdomain.org*.

Differentiated Learning

- Mathematically gifted students can explore sine and cosine functions on graphing calculators, exploring the relation between equations and the graphical wave forms produced.

- If you have access to an oscilloscope, musical students can explore the frequency of sound as digital wave forms. Without an oscilloscope, they can use iTunes to display waxing and waning volume (amplitude) as they play.

Answers to Student Questions
Part 1

1. Students moved to one side and then back all the way down the line.

2. The energy came from the force's shove.

3. The energy traveled all the way to the end of the line.

4. No, each student did not move as far as the energy traveled.

5. This was a longitudinal (or compression) wave.

6. Responses will vary depending on how vigorous "the force" was, and how responsive students were in the wave.

Part 2

7. Students in the line moved forward and backward.

8. The direction of the wave was toward the end of the line.

9. The energy from "the force" pulling on the end student made them move.

10. The energy went all the way to the end of the line.

11. This was a transverse wave.

12. Responses will vary depending on how vigorous "the force" was, and how responsive students were in the wave.

Assessment

- This is a brief exercise that leads to other Activities about waves and tides. Assessment for this Activity can be as simple as discussing it, using the questions as a launching point.
- For more formal assessment, you can grade the questions.
- If you started this Activity with a KWL as described in Preconceptions, you can do the last part (What Did You Learn?) at the end of instruction on waves.

Resources

www.ndbc.noaa.gov

www.classzone.com/books/ earth_science/terc/content/ visualizations/es1604/ es1604page01.cfm?chapter_ no=visualization

http://phet.colorado.edu

www.teachersdomain.org

http://earthquake.usgs.gov/ learn/faq

Activity 12 Planner

Activity 12 Summary

Students construct a wave tank and study the connection between wave height and wind intensity, and length of time of wind action.

Activity	Subject and Content	Objective	Materials
Waves and Wind in a Box	Wind's influence on waves	Investigate the relationship between wind and waves.	Each group will need: one or two large plastic trash bags (preferably white), 2 kg sand (optional), two sturdy cardboard boxes (75 cm × 28 cm × 5 cm or comparable size), two-speed fan or hair dryer, scissors, packing or duct tape, stopwatch or watch with second hand

Time	Key Concept	Margin Features
100 minutes	II: Ocean structure and water movement	Safety Alert!, Fast Fact, What Can I Do?, Connections, Resources

Scientific Inquiry	Unifying Concepts and Processes	Technology
Modeling, observing, and experimenting	Energy flow in open systems	Building a wave tank

Waves and Wind in a Box

Background

What are some of the ways you can make waves in water? Have you ever jumped into a pool or lake and done a "cannonball"? This makes waves. Have you ever splashed someone else when you were playing in the water? This makes waves, too. Have you ever tried blowing across the surface of water, like in a bathtub or sink? This is still another way to make waves. The tiny ripples you make by blowing across the water are waves just like the much larger waves you see in the ocean.

The largest water waves on Earth are found in the oceans, and most ocean waves are created by wind. Far out in the ocean, wind can create waves that are enormous. The largest wave ever measured accurately was 34 m high—that is about as high as a 10-story building! Some waves are big enough to capsize huge ships. How can wind make such waves? In this Activity, you will have a chance to find out.

Fast Fact

Highest sustained wind speed and/or gusts measured at sea? The strongest winds over the ocean are associated with hurricanes. The highest gust speed measured was observed off Barrow Island, Australia, during Tropical Cyclone (Hurricane) Olivia in 1996. Winds reached 103 mi./sec. (372 km/h or 230 mph).

Objective

Investigate the relationship between wind and waves.

Topic: ocean waves
Go to: *www.scilinks.org*
Code: PESO 009

Topic: tsunamis
Go to: *www.scilinks.org*
Code: PESO 010

Procedure

Part 1

1. Construct your wave tank by cutting away one end of each box and then joining the two boxes at the open end with tape. (See **Figure 12.1**.)

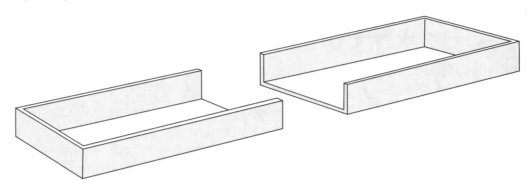

Figure 12.1
Two cardboard boxes with one end cut away

Materials

Each group will need

- one or two large plastic trash bags (preferably white)
- 2 kg sand (optional)
- two sturdy cardboard boxes (75 cm × 28 cm × 5 cm or comparable size)
- two-speed fan or hair dryer
- scissors
- packing or duct tape
- stopwatch or watch with second hand

Time

100 minutes

2. Cut a plastic bag down each side along the crease and unfold it so that it is twice as long. If this is not long enough to cover the bottom and sides of your wave tank, cut another bag and join the two with tape.

3. With your sand, construct a "beach" at one end of the tank.

4. Fill the tank with water to a depth of about 3 cm.

5. Position the fan or hair dryer so it is aimed down the tank along the surface of the water at about a 45° angle. (See **Figure 12.2**.)

Figure 12.2
Final setup, including the position of the fan (or hair dryer), the boxes and plastic bags, and the sand "beach"

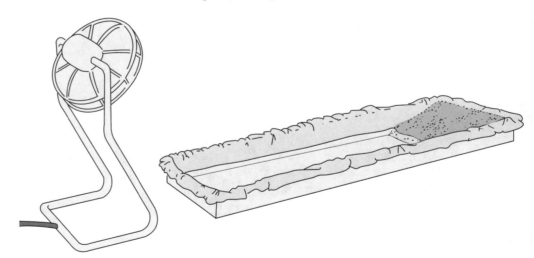

Part 2

6. You will study how waves are affected by wind. In the data table in **BLM 12.1**, write a prediction that describes how you think the speed of the fan or hair dryer will affect the waves.

7. Turn the fan or hair dryer on "low" for two minutes, and record your observations in the data table (**BLM 12.1**).

8. Turn the fan or hair dryer off and allow the water to become still. Then turn the fan or hair dryer on "high" for two minutes, and record your observations in the data table (**BLM 12.1**).

Part 3

9. Write a prediction in the data table (**BLM 12.1**) describing how you think the length of time that the fan or hair dryer blows will affect wave size.

10. Turn the fan or hair dryer on "high" for five seconds and record your observations in the data table (**BLM 12.1**).

11. Turn the fan or hair dryer off and allow the water to become still. Turn the fan or hair dryer on "high" for ten seconds and record your observations.

Questions and Conclusions

1. In Part 2, what happened to the waves when the fan or hair dryer was changed to a higher speed?

2. In Part 3, what happened to the size of the waves as time increased?

3. In Parts 2 and 3, did you notice any differences in the waves from one end of the tank to the other? If so, what were they?

4. From all your observations, what characteristics of wind are important in determining the height of a wave? Can you think of any other wind characteristic that may affect wave height?

Data Table: Wave Tank Observations

Hypothesis		Observations	
		Fan on Low	**Fan on High**
Part 1			
		Fan on High, 5 sec.	**Fan on High, 10 sec.**
Part 2			

Waves and Wind in a Box

What Is Happening?

The most common cause of ocean waves is wind. Waves caused by wind vary in height from only a few millimeters to many meters. The height of waves is determined by three factors related to wind: (1) speed of the wind, (2) the length of time the wind blows in the same direction, and (3) the horizontal distance over which the wind blows (referred to as the "fetch"). As each of these increases, the height of waves increases. In a body of water as large as the ocean, these factors can combine to produce waves of enormous height and energy. The largest wind-generated wave ever measured reliably was 34 m tall.

The wave tank used in this Activity is a very small-scale model of what happens in the ocean. Although it accurately represents the role of wind in wave formation, the small size limits the investigation of the third factor, fetch. Fetch can be explored in Activity 13 (also see Extended Learning).

It is also important to reinforce to students the difference between currents and waves, both of which can be caused by wind. Water molecules in a current move along with the current, resulting in massive amounts of water being moved on a global scale. In a wave, though, water molecules stay in basically the same place, and only the energy travels along with the wave. (Refer to Activity 11.)

Objective
Investigate the relationship between wind and waves.

Key Concept
II: Ocean structure and water movement

Materials
Each group will need

- one or two large plastic trash bags (preferably white)
- 2 kg sand (optional)
- two sturdy cardboard boxes (75 cm × 28 cm × 5 cm or comparable size)
- two-speed fan or hair dryer
- scissors
- packing or duct tape
- stopwatch or watch with second hand

Time
100 minutes

How Do We Know This?

How are scientists able to accurately measure waves in the ocean?

Scientists are able to measure ocean waves in two ways. The first way is using satellite altimeters that measure the surface of the ocean and can distinguish between the crest and the trough of a wave. The second method uses instruments set on moorings anchored in the ocean, like the ones used by the National Data Buoy Center (NDBC), or in fixed platforms. All this wave data can be assimilated into wave forecasting models that predict the ocean wave state for both coastal and open-ocean regions.

Preconceptions

If you did not do Activity 11, discuss with students what they have already learned about waves and what they would like to know. See Preconceptions in Activity 11 for specific guidance. If you did that Activity without students mentioning the effects of wind, bring it up. Dig into their knowledge about the connection between wind and waves. Ask students what they have experienced or what they have been taught formally, if anything. Some examples of student preconceptions follow:

- Wind causes all water waves.
- Waves are always largest near shore.
- Tides, currents, and boats cause waves; wind does not.
- The direction of movement of wind waves is a good indication of the local wind direction.

What Students Need to Understand

- Wind causes waves. In fact, most ocean waves are caused by wind.
- Wave height is directly related to three factors regarding wind: speed of the wind, the length of time wind blows in one direction, and horizontal distance over which wind blows, called "fetch." (Fetch is not addressed in this Activity, but it is explored in the Extended Learning feature).
- The height of waves caused by wind can range from a few centimeters to many meters.

Time Management

Construction of the wave tank and the Activity can be done together in about 50 minutes, but students may be rushed. Therefore, you might prefer to spread the two Activities over two class periods—100 minutes. Students will probably want to do some additional experimenting with the wave tank.

Preparation and Procedure

Several of the materials in this Activity can be modified. The trash bags do not have to be white, but it does help make the waves easier to observe. Inspect the bags before giving them out to students, making sure they have no holes. The sand is optional. However, having a sloped "beach" serves as a dampener to prevent waves from reflecting and interfering with the waves being generated by the fan or hair dryer. To facilitate cleanup, the sand can be put in freezer bags and molded into the shape of a beach. This prevents the sand from getting wet. If freezer bags are used, sand can be replaced by sugar, flour, or even dirt.

The cardboard boxes do not have to be of the dimensions listed. Many types can be trimmed to the desired size. A few types of boxes have been found to

SAFETY ALERT

1. Be careful to quickly wipe up any spilled water on the floor—slip and fall hazard.

2. Wash hands with soap and water upon completing the lab.

3. Use only GFI-protected circuits for appliances (e.g., hair dryer) when working with water or other liquid near electricity.

4. Be careful when working with a hot hair dryer—skin can be burned. Notify your teacher immediately if you or your lab partner is burned.

work well without major modification. One is the very shallow box in which a case of soft drinks is shipped. Another possibility is the lid of a box in which a case of paper is shipped. A third possibility is a sturdy pizza box. With any of these, a wave tank of any length can be made. The length specified in the Activity is sufficient for the purposes of this Activity, but it would be interesting to construct one that is much longer (sides of a longer tank may need reinforcement).

The fan or hair dryer is not essential. Electricity near water represents a serious safety hazard, as noted in the student section. If the fan or hair dryer comes in contact with the water, harmful and potentially fatal electric shock may occur. *Use only GFI-protected circuits.* Another option is to fan the surface of the water with a stiff piece of paper or cardboard, or wind can be generated with a balloon. Very long balloons (1 or 2 m) are available in most grocery stores. A good source of wind is made available by attaching a straw to the open end and releasing the opening of the balloon while holding onto the balloon itself.

Only one set of wave tanks is needed. The same set can be used for each class. Classes that do not make them will have more time to experiment or discuss results.

Extended Learning

- Since this Activity did not have students experiment with fetch—the relationship between wave height and horizontal distance over which the wind blows—have students investigate this relationship by constructing wave tanks of varying lengths. Students can tape three or four boxes together to make longer wave tanks. Be sure to have students compare between the different lengths. You can challenge students to devise a way to measure wave height in their wave tank. (See Demonstration in Activity 13.)

- Students can also explore the effects of fetch by monitoring wind and wave height data in real bays and oceans. The National Data Buoy Center provides wind direction (WDIR), wind speed (WSPD), and wave height (WVHT). Not all buoys provide all data. Search for "NDBC" at *www.ndbc.noaa.gov*.

- The largest wind waves are "rogue waves." Have students find more information about rogue waves: their origin, their consequences, and so on.

- Some of the largest and most destructive ocean waves are not formed by wind. They are caused by movements of Earth's crust, such as earthquakes below the ocean floor or huge landslides. These waves are called tsunamis or seismic sea waves. Encourage students to investigate the topic of tsunamis: their magnitude, their location, their frequency, and so on. (See Connections.)

Interdisciplinary Learning

- Study the culture of areas that have long depended on the oceans and seas for their existence. Japan has many accounts of tsunamis (the word is Japanese) that have reached its shores. See how tsunamis have been integrated into Japan's customs, folklore, and even artwork. There is a well-known Japanese painting entitled "The Great Wave Off Kanagawa," which shows a tsunami off the coast of Japan. You can ask students to investigate the effects of the March 11, 2011 Honshu tsunami. You can also ask them to investigate historical evidence for a tsunami in Honshu in 1700 that geologists use to date the magnitude 9 Cascadia earthquake, which originated in the Pacific Northwest.

• Have students read, interpret, and discuss Lilian Moore's poem, "Until I Saw the Sea." Students may work in pairs or groups of four to complete the following: Have students explain how the poet uses words to "paint" a picture, how the poet uses alliteration, and how she feels about life at the shore.

Differentiated Learning

• After experimenting and answering questions, some students will benefit from a video explanation of the connection between wind and waves. Teachers' Domain offers a video, "Making Big Waves" (6:18). Search for "big waves" in Teachers' Domain at *www.teachersdomain.org*.

• Windows to the Universe explains ocean waves in three reading levels in English and Spanish. Search for "windows 2 universe," then "ocean waves" at *www.windows2universe.org*.

Answers to Student Questions

1. The waves in Part 1 got bigger when the fan or hair dryer was changed to a higher speed.

2. The waves in Part 2 got bigger as time increased.

3. Yes. The waves got bigger farther away from the fan, but they were farther apart.

4. Wind speed and the length of time that the wind blows are important in determining the height of a wave. An additional factor is fetch, the length of water over which a given wind has blown. The larger the fetch, the more energy the wind can transfer to the wave.

Assessment

• While students explore the relationship between wind and waves in the lab, ask them to record the questions that arise. How would they like to vary the setup? What variables would they like to change—one at a time? What are they thinking? What are they concluding tentatively? How can they test their understanding?

• If you do not plan to do the demonstration in Activity 13, and this is the end of your instruction on waves, ask students to write what they have learned about waves. This would be the last stage of the KWL described in the Preconceptions section of Activity 11. If you will do the demonstration in Activity 13, wait to ask students to write what they have learned about waves until after you have done the demonstration.

- For formal summative assessment, you can ask students to explain their observations and their understanding of how wind affects waves in writing or cartoons; you can ask them to make a concept map about the topic; or you can grade answers to students' questions.

Resources

www.ndbc.noaa.gov

www.teachersdomain.org

www.tsunami.noaa.gov

www.windows2universe.org

Activity 13 Summary

Use a large Plexiglas tank to help students understand wave reflection and interference, the relationship between wave velocity and depth, and the effect of the seafloor on the motion of water and sand.

Activity	Subject and Content	Objectives	Materials
Tanks a Lot— Activities for a Wave Tank (Teacher Demonstration) A: What's a Wave? B: Making Waves C: Waves: The Inside Story	Modeling wave formation, properties, and behavior	A: Teach students about selected wave characteristics and properties. B: Show the effect of water depth on wave speed. C: Show the orbital motion of particles in an ocean wave, and two aspects of wave–coast interactions: breakers and sand transport.	A: water, wave tank B: water, wave tank, stopwatch C: water, wave tank, small pebbles, floating objects (Ping-Pong balls, eyedroppers, corks, etc.)

Time	Vocabulary	Key Concept	Margin Features
50 minutes	Crest, Trough	II: Ocean structure and water movement	Safety Alert!, Fast Fact, Connections, Resources

Scientific Inquiry	Unifying Concepts and Processes	Technology
Modeling, observation, measuring, and experimenting	Energy flow in open systems	Measuring waves using a wave tank

Tanks a Lot—Activities for a Wave Tank
(Teacher Demonstration)

Introduction

Activity 11 and Activity 12 help make the concepts related to waves concrete for students. They are effective because students generally have firsthand experience with waves—whether at the beach or in a bathtub.

Some concepts related to waves, though, are difficult to demonstrate in the classroom. They require observing water waves on a scale that is not ordinarily feasible for the classroom teacher because the necessary equipment is not available. For example, although it is possible to create water waves in a small space, it is difficult to observe, measure, and explore wave height, wavelength, the speed of waves and the factors that determine wave speed, and how and why waves break as they approach the shore. The teacher demonstrations presented here relate to these aspects of water waves that are normally difficult for students to observe. They were developed for a large wave tank and are designed for use in a classroom. (See **Figure 13.1**.)

Three demonstrations are presented in Activity 13: What's a Wave?, Making Waves, and Waves: The Inside Story. Included in the instructions are suggested observations and questions for class discussion. The demonstrations do not have to be done in the order presented; each can stand alone. The instructions pertaining to filling and draining the tank are written as though the tank is empty at the beginning of each demonstration.

Topic: ocean waves
Go to: *www.scilinks.org*
Code: PESO 009

Topic: tsunamis
Go to: *www.scilinks.org*
Code: PESO 010

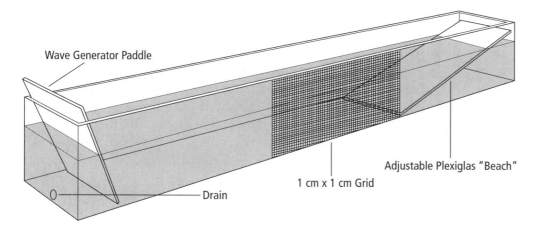

Wave Generator Paddle

1 cm x 1 cm Grid

Adjustable Plexiglas "Beach"

Drain

Figure 13.1
Wave tank

Making a wave tank is ambitious and can be daunting. If it is not something you want to take on, check to see if a science museum, college or university, oceanographic institute, or government lab in your area has one to which you can take students.

Time Management

The entire set of demonstrations can take as little as 50 minutes in one class period, or as much as a week depending on how much instruction or discussion you want to include throughout the course of the demonstrations.

Preconceptions

If you have not done Activities 11 or 12, see them for suggestions about how to uncover students' preconceptions about waves. Understanding students' preconceptions will be important if these three demonstrations are the focus of your hands-on exploration of waves—if, for instance, you are taking a field trip to use a wave tank. Students might believe the following:

- As a wave moves through matter, the matter itself moves long distances.
- Wind causes all water waves.
- Every seventh wave is a big one.
- Waves are largest near shore.
- Waves only break near shore.
- Wind causes all water waves.

Preparation and Procedure

The demonstrations described here were developed for a large, clear, Plexiglas wave tank. The tank is approximately 30 cm wide, 30 cm deep, and 270 cm (2.7 m) long. It includes three important features, two of which are attached to the bottom of the tank. One is a wave-generating paddle. The second is a hinged, 90 cm sheet of Plexiglas, used to simulate beaches of varying slope. Finally, a transparent 1 cm × 1 cm grid is included to attach to one side of the middle 90 cm section. A diagram of the wave tank is shown in **Figure 13.1**. These tanks are available commercially, or you can also build one (see Constructing a Wave Tank at the back of the book).

You should practice using the wave tank before attempting these demonstrations. Generating smooth waves requires some experience. You will also need to work out the logistics of filling and draining the wave tank. Filling the tank can be time consuming. It is important to have a hose long enough to go straight from the faucet to the tank. The tank is equipped with a drain valve, as shown in **Figure 13.1**. If the tank can be placed so the valve is over a sink, draining will be much simpler.

Fast Fact

The largest wave tank in the world is part of the O.H. Hinsdale Wave Research Laboratory at Oregon State University (see *http://wave.oregonstate.edu*). The tank is 48.8 m long and 26.5 m wide.

What's a Wave?

What Is Happening?

As discussed and investigated in Activity 11, there are two general types of waves: transverse and longitudinal. Remember that ocean waves are essentially transverse waves. (See **Figure 13.2**.) Also, remember that a longitudinal, or compression, wave can be represented as the compression and expansion of a large spring as shown in **Figure 13.3**. Sound travels in longitudinal waves.

In both types of waves, it is true that the medium through which the wave moves (water, in the case of ocean waves) has little or no net movement. Watching someone floating beyond the breakers demonstrates that while the waves move toward shore, the person bobs up and down as the waves pass. Both types of waves also have crests and troughs. The **crest** and **trough** of a transverse wave are illustrated in **Figure 13.2**. In a longitudinal wave, a region of compression can be thought of as a crest, and a region of expansion as a trough. (See **Figure 13.3**.) Waves also have wave height and wavelength. For transverse waves, wave height is defined as the vertical distance from crest to trough, and wavelength is the distance from crest to crest or trough to trough.

Waves exhibit several properties, two of which are reflection and interference. When a wave strikes an object, it reflects off the object. For example, when sound waves reflect, we may hear an echo. Light waves reflect from objects, and those entering our eyes are the ones we see. The most obvious case of this is light striking a mirror. The reflection of your image in a mirror is similar to an echo of your voice. Water waves reflect, too. This can easily be observed in a swimming pool or bathtub by watching the action of waves as they strike the sides of the pool.

Vocabulary

Crest: The highest point of a wave.

Trough: The lowest point of a wave.

Objective

Teach students about selected wave characteristics and properties.

Key Concept

II: Ocean structure and water movement

Materials

- water
- wave tank

Time

50 minutes

How Do We Know This?

How do we know that results from a wave tank provide useful information about the ocean?

The waves generated in a wave tank are similar to ocean waves in most of the important characteristics. The main differences are scale and the fact that in a tank, single waves can be isolated. The limitations of scale prevent the study of extremely large waves, as well as phenomena such as breaking under realistic conditions. The trajectory of individual particles in deepwater conditions (vanishing circular trajectories with depth) is difficult to simulate as well. The ability to isolate individual waves allows the study of wave creation and destruction without the complexity of the real world, where waves can be coming from multiple directions and have diverse characteristics (wavelength, frequency).

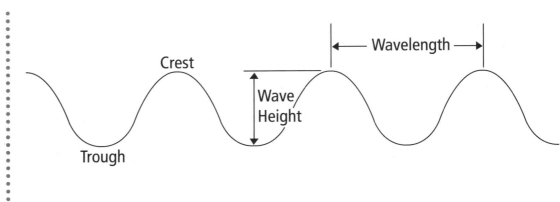

Figure 13.2
Transverse wave
and its parts

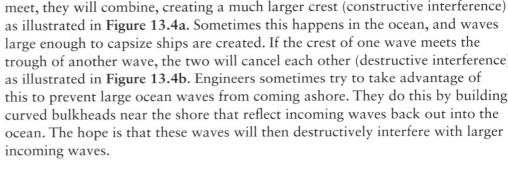

Figure 13.3
Longitudinal wave

Waves also may interfere with one another. When crests from different waves meet, they will combine, creating a much larger crest (constructive interference) as illustrated in **Figure 13.4a**. Sometimes this happens in the ocean, and waves large enough to capsize ships are created. If the crest of one wave meets the trough of another wave, the two will cancel each other (destructive interference) as illustrated in **Figure 13.4b**. Engineers sometimes try to take advantage of this to prevent large ocean waves from coming ashore. They do this by building curved bulkheads near the shore that reflect incoming waves back out into the ocean. The hope is that these waves will then destructively interfere with larger incoming waves.

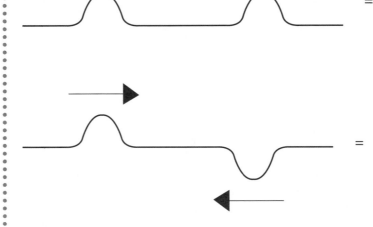

Figure 13.4a
Constructive interference

Figure 13.4b
Destructive interference

What Students Need to Understand

- Waves have crests and troughs.
- Waves reflect when they encounter an obstacle.
- Waves can interfere with each other to make larger or smaller waves.

Procedure

Note: For Demonstration A, the adjustable piece of Plexiglas—"the beach" —should be flat on the bottom of the tank. (See **Figure 13.1**.)

1. Fill the wave tank with water to a level of about 15 cm.

2. Have a student move the wave-generating paddle gently back and forth in a steady rhythm. This may take some practice.

3. Have students make observations about the waves for at least three minutes. Point out crests and troughs.

Note: You can try to measure wave height and wavelength at this point, but it is very difficult to do because of all the interference from reflected waves. It is easier to make these measurements in Demonstration C.

4. Stop generating waves. Allow the surface to become almost still.

5. With one forward motion of the wave-generating paddle, make one large wave and watch it travel the length of the tank. Ask students to observe what happens to the wave when it reaches the far end of the tank. Point out that other waves (sound and light, for example) also reflect.

6. Have a student generate waves again as in step 2. Ask students to try to pick out two crests moving toward each other from opposite ends of the tank and observe what happens when they meet. It will be easier for students to observe constructive interference if they are close enough to the tank to see the grid clearly.

SAFETY ALERT

1. Be careful to quickly wipe up any spilled water on the floor— slip and fall hazard.

2. Use only GFI-protected circuits for appliances (e.g., wave generator) when working with water or other liquid near electricity.

<table>
<tr><td>

Objective
Show the effect of water depth on wave speed.

Materials
- water
- wave tank
- stopwatch

</td></tr>
</table>

Making Waves

What Is Happening?

As a wave moves along the surface, the ocean depth affects the wave's speed. This is because a wave actually extends down below the surface of the water. The disturbance of water gradually decreases as depth increases, and below a certain depth, the wave no longer causes water movement. This depth depends on the wavelength. Generally, the depth below which there is no disturbance is about one-half of the wavelength of the wave. For example, if a wave has a wavelength of 10 m, the water is not disturbed by the wave below a depth of about 5 m. (See **Figure 13.5**.)

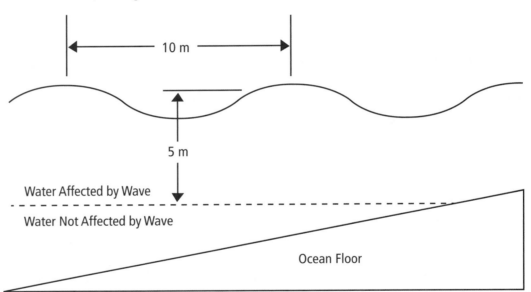

Figure 13.5
Ocean floor and wave interaction

Students will investigate the connection between wave speed and water depth in Demonstration B. You can calculate the speed of a wave by dividing the distance a crest travels by the time taken to travel it. When the water is shallower than one-half the wavelength, the ocean floor affects the speed of the wave. Friction between the water and the ocean floor slows down the wave. As the water becomes shallower, the wave slows more and more until it finally reaches the beach and stops. Sometimes, the wave will break as it approaches the beach; this effect is discussed in Demonstration C.

What Students Need to Understand

- Waves extend below the water surface. Therefore, waves are affected by underwater factors, such as depth.
- In deep water, most waves are not affected by the ocean floor.
- As the water becomes shallower near the shore—less than one-half the wavelength of the wave—the wave speed lessens due to interaction with the ocean floor.

Procedure

1. Fill the wave tank to a depth of 20 cm. The adjustable piece of Plexiglas—"the beach"—should be flat on the bottom of the tank. (See **Figure 13.1**.)

2. Ask for four volunteers: one to generate waves, one to be a timer, one to be an observer, and one to take notes.

3. After the water is almost still, let the wave generator practice sending a single wave to the other end of the tank by moving the wave-generating paddle forward once. At the same time, let the observer practice recognizing when the wave reaches the other end. (Suggestion: Waves can be difficult to see. If a light is directly above the wave tank, the arrival of a wave at the end of the tank can be recognized by the distortion of the light's reflection.)

4. After the volunteers are ready, allow the water to again become almost still. Have the wave generator say, "One, Two, Three—Go!", and on "Go," send one wave. On "Go," the timer starts the stopwatch and then stops it when the observer says, "Now," indicating that the wave has reached the other end. Record the time on the blackboard.

5. Repeat step 4 at least twice more, allowing the water to become almost still between trials. Calculate the average of the trials and have the recorder write it on the blackboard.

6. Drain the wave tank to 10 cm and repeat steps 4 and 5.

7. Drain the wave tank to 5 cm and repeat steps 4 and 5.

8. Ask students what happened to the time required for the wave to reach the far end of the tank as the depth decreased. Ask them what this implies about the speed of the wave. Ask them why wave speed decreased as the water depth decreased.

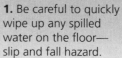

SAFETY ALERT

1. Be careful to quickly wipe up any spilled water on the floor—slip and fall hazard.

2. Use only GFI-protected circuits for appliances (e.g., wave generator) when working with water or other liquid near electricity.

Objective

Show the orbital motion of particles in an ocean wave, and two aspects of wave–coast interactions: breakers and sand transport.

Materials

- water
- wave tank
- small pebbles
- floating objects (Ping-Pong balls, eye-droppers, corks, etc.)

Waves: The Inside Story

What Is Happening?

Water in an ocean wave has little or no net motion. Water waves are described as transverse waves, and for most purposes, this description is sufficient. Strictly speaking, water waves are not transverse but are instead "surface waves." Surface waves occur at the boundary between two materials, for example, between water and air (water waves), or between the ground and air (earthquake waves). A surface wave is a combination of a transverse and a longitudinal wave. (See the What Is Happening? section of Demonstration A for descriptions of these waves.) That is, the particles in these waves move up and down and side to side, making for a nearly circular orbit. The net movement of particles in surface waves is still almost zero.

As a wave comes by, items floating in the ocean display this type of motion. If you were on a raft floating out beyond the breakers, this is what you would notice: As a crest approached, you would move backward (away from the shore) and up, then forward and up until the crest was directly underneath you. At this point, you would have completed half your circular orbit. As the crest passed, you would first move forward and down, and then down and back until you completed the orbit. This is depicted in **Figure 13.6**. The motion is very similar to what you experience on a Ferris wheel but on a much smaller scale.

Figure 13.6
Particle motion associated with a passing surface wave

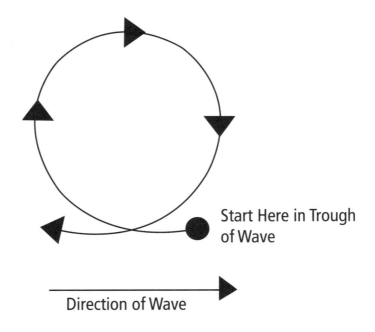

Start Here in Trough of Wave

Direction of Wave

As an ocean wave approaches the shore, the depth of the water decreases. At a depth of one-half the wavelength, the ocean floor begins to affect the wave. The first effect is a slowing of the wave. As the water becomes even shallower, the shape of the wave begins to change. This is because of the influence of the ocean floor on the orbital motion of the water particles. As the water becomes shallower, the ocean floor interferes with the motion of the water particles, and the wave begins to "pile up," as shown in **Figure 13.7**. When the ratio of the height of the wave to the depth of the water equals four-fifths, the wave becomes unstable and breaks. Not all waves break, however. Whether or not waves break is determined by several factors. Among them are the speed of the wave, the height of the wave, and the slope of the beach.

Ocean waves carry tremendous amounts of energy. You may have experienced it if you have ever been underneath a wave when it broke. The energy is dissipated in several ways. One of these is in moving sand and other things on the beach. In some cases, beaches are eroded. In other cases, they are built up.

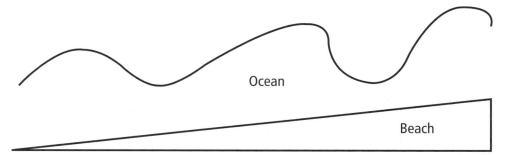

Ocean

Beach

Figure 13.7
The effect of the ocean floor on the shape of a wave

What Students Need to Understand

- Water particles in a wave are not carried along with the wave. Instead, they have little or no net movement.
- The water in a wave moves in a nearly circular pattern.
- As water becomes shallower, the ocean floor begins to affect the orbital motion of the water particles and changes the shape of the wave.
- As the wave approaches shore, its shape may become unstable enough to collapse, forming a breaker.

Procedure

1. The wave tank should be filled to a depth of 20 cm.

2. Raise the adjustable piece of Plexiglas so that it makes a 20° angle with the bottom of the wave tank. (See **Figure 13.1**.)

3. Move the wave-generating paddle in a smooth, rhythmic way. Have students make observations for at least three minutes. They can do this either aloud or in writing. Ask them what happens as the wave approaches the "beach."

SAFETY ALERT

1. Be careful to quickly wipe up any spilled water on the floor—slip and fall hazard.

2. Use only GFI-protected circuits for appliances (e.g., wave generator) when working with water or other liquid near electricity.

4. Adjust the piece of Plexiglas to increase or decrease the angle. Have students make more observations.

   Note: The "beach" acts as a wave dampener and prevents most wave reflection. Therefore, this is a good time to attempt measuring wave height and wavelength.

5. Allow the surface of the water to settle and return the beach to a 20° angle. Add several small pebbles to the beach at the water line and continue making waves. (Pebbles are recommended instead of sand because they are easier to remove from the tank.) Ask students to observe what happens to the pebbles. Ask them what they think waves would do to sand based on their observations of the wave tank.

6. While still making waves, ask students to specify in which direction the waves are traveling. They should say that the waves are traveling toward the "beach."

7. Allow the water to settle and add floating objects to the wave tank close to the side with the grid on it.

8. Generate waves and ask students to observe the motion of the floating object. Is the direction of motion the same as that of the wave?

9. Have students come up and trace with their finger or a water-based pen the path the object travels when a wave goes by. It is essential that the object be close to the grid on the side of the tank for this.

Extended Learning

- You and your students are limited only by your imagination in the number of demonstrations and student activities that you can do with a wave tank.

- You and your students can visit wave tanks in museums or at the university (Physical Oceanography, Marine Sciences, or Engineering Departments). You can also look for websites with videos showing wave tank experiments.

- Students can carefully investigate interference by making a video of the events and viewing them "frame-by-frame."

- Many types of wave action are observable in the tank. Students may want to do a more controlled experiment concerning the effect of the floor of the ocean on wave speed. They may also want to try other ways of measuring wave height and wavelength. Wave–coast interactions (particularly sand transport) are a relevant topic. Students can use the wave tank to learn why coastal barrier islands are constantly shifting.

- The topic of waves in general holds many possibilities for additional investigations. Encourage students to study the topic of sound waves and compression waves. Sound waves are used to determine water depth in oceans. Students may be interested to learn how this technique is used and how sound travels in water as compared to air.

- Another interesting distinction is deepwater waves versus shallow-water waves. This distinction is not easy to experience with the wave tank, but you can encourage students to investigate the topic. What are the differences in particle trajectory between the two cases?
- Students can also investigate the concept of standing waves as an extension to the constructive nodes part of Demonstration A.

Interdisciplinary Learning

- Have students explore the use of wave tanks in designing hulls of sailboats, ships, and barges. The Naval Surface Warfare Center, Carderock Division, operates long tanks, one at 904 m, to test the hydrodynamic performance of hulls for quiet operation, for instance, so that Navy ships will be difficult to detect. This facility near Washington, D.C., has several of these "tow tanks" with the capability of making waves as much as 0.6 m high and 12.2 m long. Middle school students use the tanks to test SeaPerch Remotely Operated Vehicles (ROVs) that they built. Ask students to explore the design of hulls, tow or wave tanks, and SeaPerch ROVs. Search *www.navsea.navy.mil/nswc/carderock/default.aspx* for "hydrodynamic tow tank3" and "MIT SeaPerch."
- Have students research Hawaiian traditional navigation skills. The ancient Hawaiians knew a lot about wave refraction and diffraction, and used that knowledge to navigate over thousands of miles of sea. People are actively reviving the practice in Hawaii. Search for "Hawaiian traditional navigation."

Assessment

- At the very least, ask students to summarize what they have learned from their work with the wave tank. They can write, draw, or tell you.
- If you did a KWL chart with students at the beginning of Activity 11, ask them to fill in the What Did I Learn? column with what they have learned about waves in general or water waves specifically.

Connections

Encourage students to investigate "planetary waves" or "atmospheric Rossby waves." What are the effects of these very long atmospheric waves on weather? The "planetary waves" are giant oscillations that create meanders in high-altitude winds and have a major influence on weather. The largest planetary waves are associated with the jet stream. The meanders at high altitude result in areas of high and low atmospheric pressure that control the weather at midlatitudes. Search for "planetary waves" at *www.nasa.gov.*

Resources

http://wave.oregonstate.edu

www.navsea.navy.mil/nswc/carderock/default.aspx

www.nasa.gov

Activity 14 Planner

Activity 14 Summary

Students plot the high and low tides for Ocracoke Island, where diurnal tides occur, for June 2011. Students are asked to probe the pattern, to apply predictions for practical uses, and to connect tides to lunar phases.

Activity	Subject and Content	Objective	Materials
Plotting Tidal Curves	Tidal patterns	Plot tide data for a period of one month and draw the tidal curve for this data.	Each student will need: pencil with eraser, red pen or a bright color crayon, scissors, clear tape, ruler

Time	Vocabulary	Key Concept	Margin Features
100 minutes	Bathymetry, Global forcing	II: Ocean structure and water movement	Fast Fact, What Can I Do?, Connections, Resources

Scientific Inquiry	Unifying Concepts and Processes	Personal/Social Perspectives	Historical Context
Graphing, analyzing data	Patterns of change within systems	Waves as a natural hazard	Historic wave events

Plotting Tidal Curves

Background

Anyone who has been to the coast realizes there is a rhythm to the ocean. Waves crash onto the beach or rocks. The water they carried washes ashore, then retreats. Another wave crashes ashore. The pattern repeats. However, there is another rhythmic phenomenon along the coast—tides. Tides are created by the gravitational forces between the Moon, the Sun, and Earth. In the case of lunar tides, the bulges on opposite sides of Earth are caused primarily by two factors: the gravitational pull of the Moon, and the inertia of water in the ocean.

Tides are predictable changes in sea level that occur at regular intervals. There is a high tide when sea level has risen to its highest point, and at low tide sea level has dropped to its lowest point. These changes affect humans in many ways. A person who falls asleep on the beach at low tide may wake up drenched if he or she does not move before high tide comes. One of the most important impacts of tides is on ocean shipping. In many locations, ships can only come to shore at high tide. If they come in at low tide, they may become grounded.

Since tides affect us in so many ways, it is important to know when they will occur. As already mentioned, it is possible to predict when tides will occur. By making a graph of the time of the tides and the sea level, you can begin to see the pattern of changes over a period of time. Although this can be done for any location on the coast, the location used in this Activity is Ocracoke Island on the Outer Banks of North Carolina.

Fast Fact

Tides have been predicted accurately long before computers. The first documented tide table was created in China in 1056 AD for the Qiantang River. The first systematic tidal prediction was made by William Thomson (Lord Kelvin) in 1867. He constructed a tide-predicting machine that added six tidal components by using pulleys. The pulley system could be adjusted to change amplitude and phase. When do you think we stopped using tidal prediction machines similar to Lord Kelvin's?

SC**L**INKS
THE WORLD'S A CLICK AWAY

Topic: tides
Go to: *www.scilinks.org*
Code: PESO 011

Objective

Plot tide data for a period of one month and draw the tidal curve for this data.

Materials

Each student will need

- pencil with eraser
- red pen or a bright color crayon
- scissors
- clear tape
- ruler

Time

100 minutes

What Can I Do?

You can use your knowledge to plan a trip to the beach. Where would you love to go? Who do you want to take with you? Use tide charts to pick the best couple of days in the summer to fit what you want to do. For instance, do you need a wide beach for volleyball? Would you like the tide to come in over a hot beach so that swimming will be comfortable? Do you need high tide to get a boat in the water easily? Look at tide tables for the area in local beach stores, or search online for "NOAA Tide Tables" at *http://tidesandcurrents.noaa.gov/tide_predictions.shtml*.

Procedure

1. Refer to **Table 14.1** and **BLM 14.1** titled "June 2011—Ocracoke Island, N.C. Tidal Prediction Data Sheet." Time is given in the military convention (or 24-hour time). Until 1200, or 12:00 noon, the two conventions are the same. After that, just subtract 1200 from the military time to get the civilian time. For example, 1430 means that it is 2:30 p.m. On the graphing charts, 6 corresponds to 0600 in the time column of the "June 2011—Ocracoke Island, NC Tidal Prediction Data Sheet."

2. Plot the data on the graphing chart (**BLM 14.1**) with your pencil. It is important that you plot the points with the pencil so that if you make mistakes, you can change them. After you have plotted all the data, go back and check for correctness.

3. With your pencil and ruler, connect all the points in the order they were plotted. Check for correctness.

4. With your pen or crayon, go over the pencil lines. This will make them much easier to see.

5. Cut out the three charts. Along the bottom of the charts, you will notice some circles on days 1, 8, 15, and 23 that represent the phases of the Moon. Be sure not to cut these out when you cut the charts.

6. Tape the three charts together so they make one continuous chart.

Questions and Conclusions

1. About how much time passes between one high tide and the next high tide? From one low tide to the next low tide?

2. When did the *highest* high tide occur? The *lowest* low tide?

3. When did the *lowest* high tide occur? The *highest* low tide?

4. You are staying at a beach house. At high tide, the water completely covers the part of the beach that is usable. On June 9, you go out on the beach at 10:00 a.m. to read a book. Will you be able to find a dry place to sit down? After an hour, the sound of the waves lulls you to sleep. How long can you sleep before you must either wake up or get wet?

5. Ocracoke Island has shallow sand bars all around it that will prevent even small sailboats from coming close to shore at low tide without being grounded. However, at high tide, these sand bars do not cause a problem for small boats. On June 20, the crew of a sailboat wants to get as close as possible to the island without getting stuck on the sand bars. They radio you at the Coast Guard station to ask what time they should come in to take advantage of high tide. What should you tell them? (Give the time in civilian convention.)

Table 14.1: June 2011—
Ocracoke Island, N.C., Tidal Prediction Data Sheet

Date	Time	Sea Level (feet)	High/ Low	Date	Time	Sea Level (feet)	High/ Low
01/06/2011	2:55	0.15	L	08/06/2011	19:29	0.14	L
01/06/2011	7:44	0.79	H	09/06/2011	1:45	1.02	H
01/06/2011	13:26	0.09	L	09/06/2011	8:00	0.03	L
01/06/2011	20:14	1.17	H	09/06/2011	14:30	1.07	H
02/06/2011	3:28	0.14	L	09/06/2011	20:58	0.16	L
02/06/2011	8:26	0.80	H	10/06/2011	2:43	0.96	H
02/06/2011	14:04	0.07	L	10/06/2011	8:58	0.01	L
02/06/2011	20:55	1.20	H	10/06/2011	15:33	1.14	H
03/06/2011	3:55	0.13	L	10/06/2011	22:32	0.14	L
03/06/2011	9:09	0.82	H	11/06/2011	3:44	0.91	H
03/06/2011	14:47	0.05	L	11/06/2011	10:00	-0.01	L
03/06/2011	21:38	1.21	H	11/06/2011	16:36	1.21	H
04/06/2011	4:21	0.12	L	11/06/2011	23:45	0.10	L
04/06/2011	9:55	0.84	H	12/06/2011	4:46	0.88	H
04/06/2011	15:32	0.04	L	12/06/2011	11:04	-0.03	L
04/06/2011	22:22	1.20	H	12/06/2011	17:36	1.26	H
05/06/2011	4:53	0.10	L	13/06/2011	0:47	0.06	L
05/06/2011	10:43	0.87	H	13/06/2011	5:47	0.87	H
05/06/2011	16:22	0.05	L	13/06/2011	12:06	-0.05	L
05/06/2011	23:09	1.18	H	13/06/2011	18:34	1.30	H
06/06/2011	5:32	0.08	L	14/06/2011	1:43	0.03	L
06/06/2011	11:35	0.91	H	14/06/2011	6:45	0.87	H
06/06/2011	17:17	0.08	L	14/06/2011	13:05	-0.06	L
06/06/2011	23:58	1.14	H	14/06/2011	19:28	1.31	H
07/06/2011	6:17	0.07	L	15/06/2011	2:36	0.01	L
07/06/2011	12:30	0.96	H	15/06/2011	7:41	0.87	H
07/06/2011	18:18	0.11	L	15/06/2011	14:01	-0.05	L
08/06/2011	0:50	1.08	H	15/06/2011	20:20	1.30	H
08/06/2011	7:06	0.05	L	16/06/2011	3:27	0.00	L
08/06/2011	13:29	1.01	H	16/06/2011	8:34	0.88	H

Note: The date is given as day/month/year.

Table 14.1: June 2011—
Ocracoke Island, N.C., Tidal Prediction Data Sheet (cont'd)

Date	Time	Sea Level (feet)	High/Low	Date	Time	Sea Level (feet)	High/Low
16/06/2011	14:54	-0.03	L	23/06/2011	21:34	0.28	L
16/06/2011	21:08	1.27	H	24/06/2011	2:17	0.81	H
17/06/2011	4:16	0.01	L	24/06/2011	9:08	0.20	L
17/06/2011	9:24	0.88	H	24/06/2011	15:06	0.93	H
17/06/2011	15:46	0.01	L	24/06/2011	22:34	0.28	L
17/06/2011	21:54	1.22	H	25/06/2011	3:06	0.76	H
18/06/2011	5:03	0.03	L	25/06/2011	9:28	0.19	L
18/06/2011	10:13	0.87	H	25/06/2011	15:57	0.97	H
18/06/2011	16:36	0.06	L	25/06/2011	23:30	0.26	L
18/06/2011	22:39	1.16	H	26/06/2011	3:56	0.74	H
19/06/2011	5:50	0.06	L	26/06/2011	10:01	0.18	L
19/06/2011	11:01	0.87	H	26/06/2011	16:47	1.01	H
19/06/2011	17:27	0.12	L	27/06/2011	0:21	0.24	L
19/06/2011	23:21	1.09	H	27/06/2011	4:47	0.73	H
20/06/2011	6:35	0.09	L	27/06/2011	10:43	0.16	L
20/06/2011	11:48	0.87	H	27/06/2011	17:35	1.06	H
20/06/2011	18:22	0.17	L	28/06/2011	1:07	0.22	L
21/06/2011	0:03	1.01	H	28/06/2011	5:37	0.74	H
21/06/2011	7:18	0.13	L	28/06/2011	11:27	0.13	L
21/06/2011	12:36	0.87	H	28/06/2011	18:21	1.11	H
21/06/2011	19:23	0.22	L	29/06/2011	1:48	0.19	L
22/06/2011	0:46	0.93	H	29/06/2011	6:25	0.76	H
22/06/2011	8:00	0.16	L	29/06/2011	12:13	0.09	L
22/06/2011	13:24	0.88	H	29/06/2011	19:05	1.16	H
22/06/2011	20:29	0.26	L	30/06/2011	2:26	0.16	L
23/06/2011	1:30	0.86	H	30/06/2011	7:13	0.79	H
23/06/2011	8:38	0.18	L	30/06/2011	12:59	0.06	L
23/06/2011	14:14	0.90	H	30/06/2011	19:49	1.20	H

Note: The date is given as day/month/year.

Activity 14: Plotting Tidal Curves

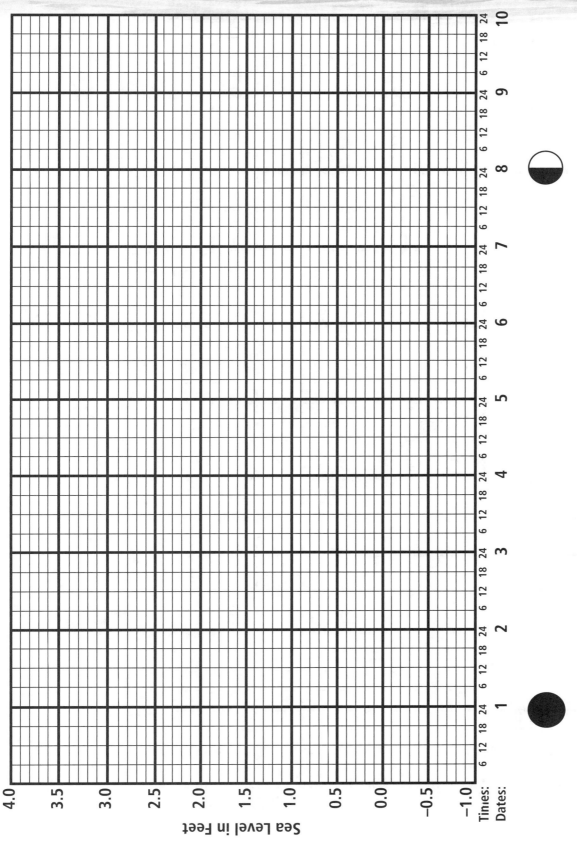

Sea Level in Feet

Times:

Dates:

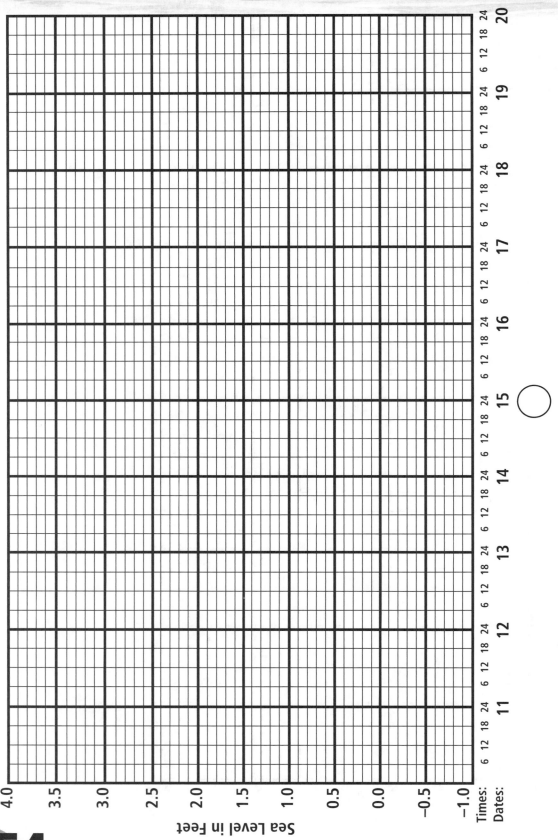

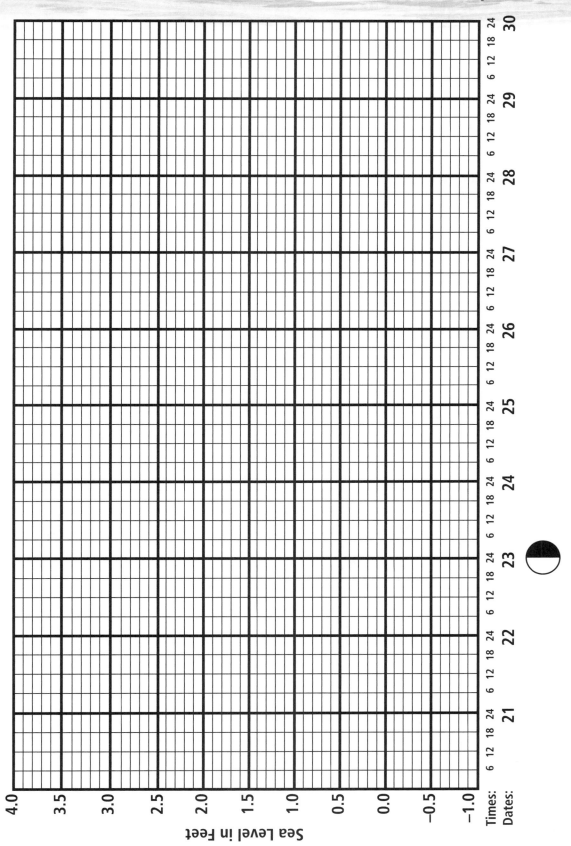

Plotting Tidal Curves

What Is Happening?

Tides are experienced along the coastlines of the world as rhythmic fluctuations in sea level. They occur predictably at regular intervals. The pattern of tides, however, is different for every location. This variation from place to place in the pattern of tides is due to a number of factors, the most important being the nature of the land. By plotting the tidal data for a specific location, a picture of the pattern of tides for that place can be obtained and predictions for future tides can be made.

The reasons for tides are not discussed in this Activity. They are discussed in Activity 15 and Activity 16. Spring and neap tides are not discussed here either but are included in Activity 15. Questions 2 and 3 in the student section of this Activity lead naturally into a discussion of spring and neap tides.

How Do We Know This?

How do we predict tides so accurately, given how many variables there are?

Old tide charts were a good approximation of tidal elevations; however, as scientists have identified more than 30 tidal constituents, accurate tidal prediction requires taking into account all those constituents, and some of them are extremely local. Model simulations of the tide provide a good approximation of the tidal elevation by including local **bathymetry** and **global forcing**. Even though scientists understand the mechanisms controlling tides, the best tidal predictions still partly remain an art and require adapting predictions to many factors, such as changes in the positions of sandbars or long-term water-level changes. For this reason, NOAA continues to update their tidal predictions periodically, and they often modify their local references. See *http://tidesandcurrents.noaa.gov/restles6.html*.

Vocabulary

Bathymetry: Measuring the depth of the ocean bottom; from the Greek *bathus* for deep, and *metrus* for measure.

Global forcing: The tide-generating force acting on the entire Earth's ocean.

Preconceptions

Find out from students if they have any funny or scary tide-related stories to share. Then ask them about patterns of tides. Do high and low tides have any patterns or do high and low water occur randomly? What is their experience, and what is their understanding of tidal patterns? Students may say the following:

- Tides are unpredictable.
- Tides occur like clockwork.
- Spring tides occur only in the spring.
- "Tidal waves" are related to tides.
- Tides are the same everywhere.

Objective

To plot tide data for a period of one month and draw the tidal curve for this data.

Key Concept

II: Ocean structure and water movement

Materials

Each student will need

- pencil with eraser
- red pen or a bright color crayon
- scissors
- clear tape
- ruler

Time

100 minutes

What Students Need to Understand

- Rhythmic fluctuations in sea level are called tides.
- Tides occur at regular intervals. As a result, time and height of tides are predictable.
- The pattern of tides varies from location to location.
- Knowing tidal patterns is important to people who seek livelihoods or recreation from the oceans.

Time Management

This Activity should take about 100 minutes or two class periods. It will take students more than 50 minutes to plot the data. To save considerable time, have students plot the data with a spreadsheet or mathematical graphing software.

Preparation and Procedure

Very little preparation is required for this Activity. Note that feet are used, not meters.

To download the data for plotting with a spreadsheet, go to *http://tidesandcurrents.noaa.gov* and search for "NOAA tide predictions 8654792." (The number gives you Ocracoke data.) Select the starting date of June 1, 2011, and choose monthly for the Time Range.

Extended Learning

- The pattern of tides that students will see in this Activity is called semidiurnal, which means two high tides and two low tides each day. Have students investigate other tidal patterns, identifying factors that account for the different patterns.
- Students often have the misconception that a tidal wave is caused by tides. "Tidal wave" is actually a misnomer. These waves have nothing to do with tides. The correct term is *seismic sea wave* or *tsunami*. Students might investigate this phenomenon and learn why it is truly not a tidal wave.

Interdisciplinary Study

- The change in sea level could be used to generate electricity. This is being done in several places around the world. What are the factors that make a location on the coast suitable for such a project? What factors might limit the use of this technology?

- Ask students to read aloud Henry Wadsworth Longfellow's poem, "The Tide Rises, the Tide Falls," to a partner or a small group. How does it sound? What techniques did Longfellow use to give rhythm to the poem? What do students think is the point of the traveler?
- This Activity applies the mathematical skill of graphing. Consider enlisting the help of students' mathematics teachers in recognizing and describing the mathematical patterns shown by tides. Or, can you collaborate with math teachers on this lesson?

Differentiated Learning

Some students might have difficulty graphing the negative numbers. If so, show them how to graph them, pair or group them with patient students who can explain how to do it, or have them compare their handmade graph to one made with a computer graphing program. You could also show them an example from the website with Ocracoke data. See the Preparation and Procedure section. The website shows the data as both a table and a graph.

Answers to Student Questions

1. Between high tides: about twelve and a half hours; between low tides: about twelve and a half hours

2. Highest high tide: June 14; lowest low tide: June 14

3. Lowest high tide: June 27; highest low tide: June 23 and 24

4. Yes, you will be able to find a dry place to sit down because it is low tide. High tide occurs at 2:30 p.m., so you have about two or three hours at the most to sleep before moving or getting wet.

5. The sailboat should come in close to 11:48 a.m. to take advantage of high tide.

Assessment

- While students are graphing, circulate through the room to make sure that their graphs are reasonable. If they are not, you will need to instruct them in plotting points. See Differentiated Learning for ideas.
- For summative assessment, you can give students data from another location, ask them to graph it, and compare and contrast the tidal patterns.
- You can also grade students' answers to the questions.

Connections

Solid Earth tides are caused by the same principles as ocean tides. The difference is that while ocean tides affect only the ocean, Earth tides affect the entire volume of Earth. The predominant solid Earth tide has a period of over 12 hours and a magnitude of around 0.5 m. The Earth tides have an impact for GPS calculations and particle physics experiments, and they have a significant effect on Earth's magnetic field changes.

Resources

http://tidesandcurrents.noaa. gov/tide_predictions.shtml

http://tidesandcurrents.noaa. gov/restles6.html

http://tidesandcurrents.noaa. gov

Activity 15 Planner

Activity 15 Summary

Students build a mobile as a model of the Earth–Moon-Sun system. By changing the relative positions of the Earth, Moon, and Sun, they explore the causes of tides.

Activity	Subject and Content	Objective	Materials
Tides Mobile	Why tides form	Construct a mobile that shows the relationship among the Sun, Moon, and Earth, and use this mobile to investigate how tides are created.	Each group will need: scissors, coat hanger, string, meter stick or dowel, modeling clay, tape, yellow construction paper, pencil, paper clip

Time	Vocabulary	Key Concept	Margin Features
50 minutes	Gravitational forces, spring tides, neap tides	II: Ocean structure and water movement	Safety Alert!, Fast Fact, What Can I Do?, Connections, Resources

Scientific Inquiry	Unifying Concepts and Processes	Technology
Modeling and explaining	Modeling to explain patterns of change within systems	Building a planet/tide model

Tides Mobile

Background

The Sun, Moon, and Earth are three extremely large objects separated by great distances. (The Moon and Earth are 384,000 km apart, and the Sun and Earth are 150 million km apart!) Despite the large distances between them, each object affects the others. Earth is kept in orbit around the Sun by the **gravitational forces** between them. The Moon is kept in orbit around Earth by the gravitational forces between them. These gravitational forces are mutual, meaning each object attracts *and* is attracted by the other.

Gravitational forces between the Moon and Earth create the lunar tides. As you investigated in Activity 14, the coastlines of Earth experience tides as rhythmic fluctuations in sea level. Lunar tides are caused primarily by two factors: the gravitational pull of the Moon, and the inertia of water in the ocean on Earth. Gravity creates a bulge of water on the side of Earth facing the Moon, which we will explore in this Activity. In fact, it is the difference on the horizontal component of gravity that drives the water toward the bulge. (Inertia creates a bulge on the side of Earth facing away from the Moon and will be explored in Activity 16.)

The Sun also creates tides. It generates two much smaller bulges—so small that, most of the time, they are not even noticed. As with the Moon, one bulge is caused by gravity and always faces the Sun. The other is caused by inertia and is always on the side facing away from the Sun. However, four times a month, the Sun's effect on Earth's tides is noticeable.

Vocabulary

Gravitational forces: The *mutual* attraction between two objects.

Spring tides: Twice-a-month tides that are higher than usual.

Neap tides: Twice-a-month tides that result in a lower high tide and a higher low tide.

Fast Fact

The wave made by tides has such a long wavelength that it is affected by the coast everywhere in the world (tidal wavelength in the deep ocean is about 10,000 km). The maximum tidal bulge on the open ocean is about 1.5 m, but it increases rapidly toward the coast. A good idea of what the global tides look like can be seen using ocean models (e.g., see *http://volkov.oce.orst.edu/tides*).

Topic: tides
Go to: *www.scilinks.org*
Code: PESO 011

Objective

Construct a mobile that shows the relationship among the Sun, Moon, and Earth, and use this mobile to investigate how tides are created.

Activity 15

Materials

Each group will need

- scissors
- coat hanger
- string
- meter stick or dowel
- modeling clay
- tape
- yellow construction paper
- pencil
- paper clip

Time

50 minutes

Figure 15.1
This mobile shows the positions of the "Moon," "Earth," "Sun," and counterbalance weight. Earth and the Moon are shown in approximately correct relative sizes, but the Sun would have to be much larger to be of correct relative size. The distances between Earth–Moon and Earth-Sun are not drawn to scale.

Twice a month, the Sun, Moon, and Earth align to produce very high and very low tides, called **spring tides**. Also twice a month, the three are aligned in a way that produces moderate tides, called **neap tides**.

To see how the Sun, Moon, and Earth interact to create tides, you will build a model of the three objects to illustrate the relative positions between them and how these positions change over the course of a month.

Procedure

Part 1

1. Begin by cutting a replica of the Sun out of construction paper. (See **Figure 15.1**.) Tape this to the end of a piece of string, and then tie the string to the middle of the meter stick (or dowel).

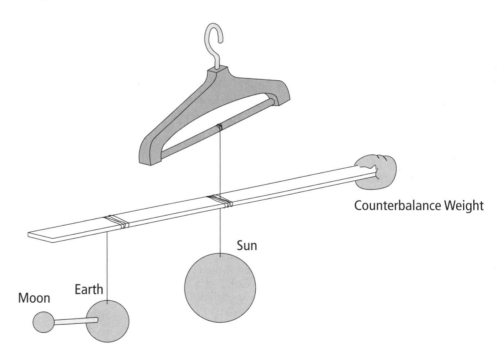

2. Make two balls of clay, one much smaller than the other. They represent the Moon and Earth. Attach a ball to each end of a pencil. (See **Figure 15.3** for relative size and placement.)

3. Place a partially unfolded paper clip (see **Figure 15.2**) in the larger ball of clay as shown in **Figure 15.3**. Tie a piece of string to the paper clip after it has been inserted in the clay. Hold the string by the free end. If the pencil is not level, adjust the sizes of the clay balls until it is. Note: It is important to realize that the Moon and Earth are "balanced" the same way as the two

162

clay balls you are trying to balance. The situation is like a seesaw; a heavier person must sit closer to the balance point than a lighter person for the two objects to balance. Earth is so much "heavier" than the Moon (just like objects in the clay models) that the balance point is located inside Earth. Earth and the Moon rotate around this common point just like the clay models do.

4. Tie the string to the meter stick at a point two pencil lengths from the string attached to the Sun. Adjust the level of the string so that the Moon and Earth are at the same level as the Sun.

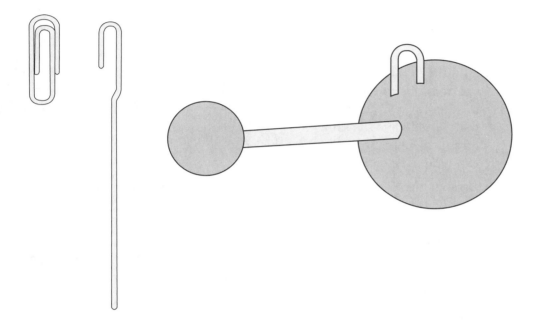

Figure 15.2 (far left)
Unfold a paper clip as shown.

Figure 15.3
Place the paper clip inside the "Earth" clay ball, as shown.

5. Tie one end of another piece of string to the center of the meter stick. Tie the other end to the center of the horizontal piece of the coat hanger. (See **Figure 15.1**.)

6. On the opposite end of the meter stick from the Moon and Earth, attach enough weight (clay, for example) so that when the mobile is held by the coat hanger, the meter stick is level, as shown in **Figure 15.1**.

Part 2

7. Once you have built the mobile, gently push the "Moon" so that the Earth–Moon system rotates. Then gently push one end of the meter stick in the same direction so that the whole mobile revolves around the Sun. This is a model of how the Sun, Moon, and Earth move relative to each other. (Obviously, though, the size and distance of your model are not to scale.)

8. **Figure 15.4** shows the part of the model with just the "Moon" and "Earth." With your pencil, shade in where the tidal bulges would be in this model.

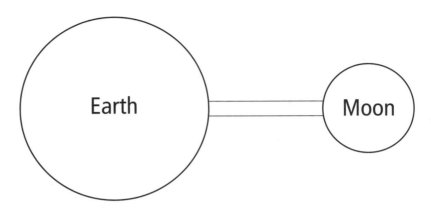

Figure 15.4
Earth–Moon part of the mobile. Shade the proper position of the tidal bulges.

What Can I Do?

If you live near a coast, keep an eye on the tides and phases of the Moon. How high does the water rise? How low does it go? What is the maximum tidal range during a year? How does that affect recreation and work for people whose lives are tied to the water? For instance, are there times when clam digging is best? If you don't live near a beach, you can watch tides on webcams, like that at the Monterey Bay Aquarium. Search for "Monterey Bay Aquarium" and look under Animals & Activities at *www. montereybayaquarium.org*.

9. Stop the mobile from rotating. Arrange the Sun, Moon, and Earth so that the tides due to the moon and the tides due to the Sun add to each other, making for a larger-than-usual tide. Is there another possible arrangement that would have the same or a very similar effect? Draw all the possible arrangements that would have this effect in the space below and explain why you have chosen them.

10. Arrange the Sun, Moon, and Earth so that the tides due to the Moon and the tides due to the Sun "subtract" from each other. That is, in what arrangement would the tides due to the Sun and the Moon pull water away from each other? Is there another possible arrangement that would have the same or a very similar effect? In your notebook, draw all the possible arrangements that would have this effect and explain why you have chosen them.

Tides Mobile

What Is Happening?

The relationships among the Sun, Moon, and Earth have many effects on Earth. One of the most apparent is the formation of tides. Along the coastlines of the planet, tides are experienced as rhythmic fluctuations in sea level. While many factors act to create tides, two are most important: gravity and inertia.

Lunar tides are due to the relationship between the Moon and Earth. The Moon's gravity pulls the entire Earth toward it. Since water is a fluid, and the rock and minerals that make up Earth are solids, the water is pulled farther than the solids. This creates a bulge of water on the side of Earth facing the Moon. (The atmosphere also experiences "tides," but they are much more difficult to measure and do not affect us as much as the water-based tides.) The bulge on the other side of Earth is due to inertia.

One of the key points of this Activity is to demonstrate to students that the Moon and Earth form a two-body system that rotates around a point. This point, however, is actually located within Earth at the center of the mass of the two bodies. An effect of this rotation is that water tends to move, because of inertia, toward the side of Earth away from the Moon, creating another bulge. (This aspect of tidal formation is addressed in greater detail in Activity 16.)

These same factors—gravity and inertia—act between the Sun and Earth, but to a lesser degree. Approximately two-thirds of the tides are caused by the Moon-Earth relation, and one-third by the Sun-Earth interaction. Two much smaller bulges are created: one on the Sun-side of Earth, and one on the other side. Normally, these bulges are not as noticeable within the tides. However, when the Sun, Moon, and Earth are positioned to form either a straight line or a right angle, the effect on the tides caused by the Sun is noticeable. When the three form a straight line—at new Moon and full Moon—the bulges due

Objective

Construct a mobile that shows the relationship among the Sun, Moon, and Earth, and use this mobile to investigate how tides are created.

Key Concept

II: Ocean structure and water movement

Materials

Each group will need

- scissors
- coat hanger
- string
- meter stick or dowel
- modeling clay
- tape
- yellow construction paper
- pencil
- paper clip

Time

50 minutes

How Do We Know This?

How do scientists measure tides?

Scientists have been measuring tides for many years, and currently the most common way of measuring tides is practically unchanged since the early 1800s. The measurement is taken in a well that is connected with the ocean, but enclosed enough to remove wave action. Nowadays, the measurement is done acoustically and the recording is done electronically, but the method remains unchanged. Additionally, global tides are measured using altimeters on board satellites like *TOPEX/Poseidon*.

to the Sun enhance those due to the Moon, creating very high and very low tides. These are called spring tides. When the three form a right angle—at first and third quarter—the bulges due to the Sun detract from those due to the Moon, creating moderate high and low tides. These are called neap tides.

Preconceptions

If you did Activity 14, you already know what students understand about patterns of tides. That Activity also points out a relationship between the timing of tides and lunar phases. Ask students, "What do you understand about how the Moon causes tides?" Then be prepared to hear some of the preconceptions listed below. If you did not do Activity 14, you will need to broaden the question to "What can you tell me about tides? What patterns do they follow? What can you tell me about their causes?" The following are possible preconceptions students might have:

- Tides are unpredictable.
- The Moon revolves around the center of Earth.
- The Moon by itself is the cause of our tides.
- There is only a tidal bulge on the Moon's side of Earth.
- Spring tides take place once a year during spring.
- Strong tidal currents occur when the tide is highest.
- High tides in nearby locations always occur at the same time.

What Students Need to Understand

- The factors that are responsible for tides are due to the positional and gravitational relationship between the Sun, Moon, and Earth.
- The two factors primarily responsible for tides are gravity and inertia. These factors operate in the Earth–Moon system and the Earth-Sun system.
- Earth and the Moon form a two-body system that rotates about a point that is located within Earth.
- The tidal bulges due to the relationship between the Sun and Earth are not nearly as noticeable as those due to the relationship between the Moon and Earth.
- The tidal bulges due to the Sun are most noticeable when the Sun, Moon, and Earth are positioned to form either a straight line (spring tides), or a right angle (neap tides).

Time Management

Students can do this Activity in 50 minutes, including the time they need to build the mobile. You might leave the mobiles up throughout your instruction on tides.

Preparation and Procedure

Be sure that all materials are centrally located or already distributed to student groups. A coat hanger will work for this Activity. Some dry cleaners will give these away. If meter sticks are not available, you can buy dowels at most hardware stores. Modeling clay is sold at most craft stores. It may also be available from the art department.

Extended Learning

- Tidal patterns vary around the world with regard to height and number of high and low tides per day. Several factors are responsible for this. The Bay of Fundy, for example has tides that are regularly 12 m high, while most locations have high tides of only 1 m or so. Have students investigate why such variations occur.

- Tides affect marine life in many ways. Many marine organisms time their spawning to benefit from specific tidal conditions. Several fish species use tides to improve the chances of their larvae to get to optimal growing conditions, like getting into estuaries. Sea turtles only come ashore to lay their eggs during spring tides. Encourage students to find out why this is so. They can also explore how tides lead to zones of organisms, such as seaweeds. See Differentiated Learning for a short video about life in intertidal zones.

- The worst time for a hurricane to make landfall is during high tide. The hurricane's storm surge adds to the tide's high water. Have students investigate the impact of high tides with hurricanes on coastal communities. They can look for examples of hurricanes that landed at the best of times and at the worst of times, and their relative destruction.

Interdisciplinary Study

- Planning for the invasion of Normandy on D-Day in 1945 included knowing tides and phases of the Moon. Landing forces needed low tide to see obstacles shaped like giant jacks ("hedgehogs") that Germans had placed in harbors. Paratroopers needed a full Moon for visibility. Ask students to explore Earth science connections with D-Day. For a firsthand account documented by the Library of Congress, search for "Planning D-Day LOC" at *www.loc.gov/loc/lcib/0304/rubber-map.html*.

- Have students examine historical maps of Boston Harbor to imagine how 18th century life depended on knowing the tides. The Library of Congress has maps and documents online about General Washington's efforts against British troops in the Boston area in 1776. The link also leads to panoramic images of Boston through the years, showing the evolution of the harbor. Search for "Memory Washington March 24" at *http://memory.loc.gov/ammem/today/mar24.html*.

Connections

The same way that the Moon causes alterations in the movement of Earth (and in reality we have a two-body system rotating around a center of mass), Earth and the rest of the planets in the solar system cause alterations in the movement of the Sun. This fact has a lot of applications for the discovery of extrasolar planets. The method is called astrometry and consists of carefully measuring the position of a star and its temporal evolution. If that star has a planet orbiting it, then they will rotate as a two-body system. As the planet is much smaller than the star, we may not see the planet, but its effects on the motion of the star can be detected and the existence of the planet can be postulated. Astrometric techniques are still under development for extrasolar planetary detection, but they have been extremely useful to the study of binary star systems. Search for "Extrasolar Techniques" at *www.esa.int*.

Differentiated Learning

- For students with little firsthand experience with tides and salt marshes, Teachers' Domain at *www.teachersdomain.org* provides vivid images of flora and fauna living in salt marshes along Cape Cod. "Intertidal Zone" comes from NOVA's "The Sea Behind the Dunes." This 4:52 video has audio and captions.

- Windows to the Universe explains tides in three reading levels in English and Spanish. Search for "Windows 2 Universe" and then "Tides of the Ocean" at *www.windows2universe.org/earth/Water/ocean_tides.html*.

Answers to Student Questions in Procedure

8. Tides should be indicated as shown in **Figure 15.5**.

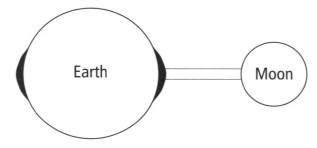

Figure 15.5
This illustration shows how to indicate tides.

9. One possible arrangement is shown in **Figure 15.6**.

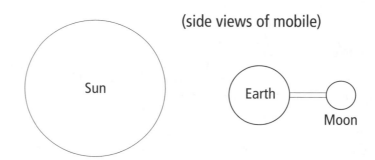

(side views of mobile)

Figure 15.6
This illustration shows one possible arrangement that would have the same or a very similar effect.

It would make the tides very high and very low because the bulges due to the Sun would add to the bulges due to the Moon. Another possible arrangement is shown in **Figure 15.7**. The tides would be higher and lower than usual for the same reason as above: the bulges due to the Sun would add to the bulges due to the Moon. Both of these arrangements create spring tides.

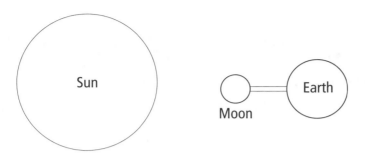

10. Both possible arrangements are shown in **Figure 15.8**. In both cases, the alignment would make moderate high and low tides because the bulges due to the Sun would detract from the bulges due to the Moon. These arrangements create neap tides. (Note: The following views are from above the mobile. Since a piece of construction paper is used to represent the Sun in this Activity, the "Sun" is represented by a straight line in the two drawings.)

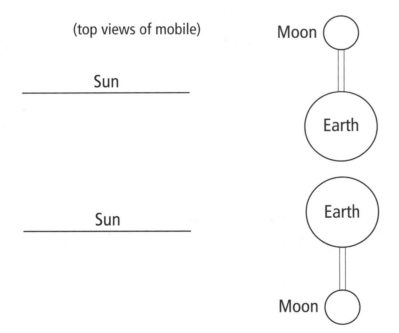

(top views of mobile)

Figure 15.8
This illustration shows both possible arrangements, which create neap tides.

Assessment

- Ask students to draw cartoons showing the position of the Earth–Moon-Sun system at times of neap and spring tides. The cartoons should also show the bulges of water representing high tides.
- You could also ask students to create a video using their model to explain tides to someone who is not in their class. They would need to write a script for the video rather than improvise their explanation. You can have a class discussion about the qualities of a good video, based on what they see. Afterward, you can give them a chance to redo the video.
- You could also grade questions 8 to 10 for completeness and accuracy.

Resources

http://volkov.oce.orst.edu/tides

www.montereybayaquarium.org

www.loc.gov/loc/lcib/0304/rubber-map.html

http://memory.loc.gov/ammem/today/mar24.html

www.teachersdomain.org

www.windows2universe.org/earth/Water/ocean_tides.html

www.esa.int

Activity 16 Planner

Activity 16 Summary

Students build a model of the Earth–Moon system to which they attach masses on strings. With this model, they explore the role of inertia in causing tides.

Activity	Subject and Content	Objective	Materials
The Bulge on the Other Side of Earth	Inertia as a factor in tides	Demonstrate how the rotation of the Earth–Moon system accounts for the bulge of water on the side of Earth facing away from the Moon.	Each group will need: safety glasses or goggles for each student, Styrofoam ball (15 cm diameter), Styrofoam ball (6 cm diameter), string, dowel (1 m long and 0.5 cm diameter), dowel (0.5 m long and 0.5 cm diameter), masking tape, two weights (about 15 g each), 473 ml (16 oz.) drink bottle, ruler

Time	Vocabulary	Key Concept	Margin Features
50 minutes	Inertia, Axis of rotation, Center of gravity	II: Ocean structure and water movement	Safety Alert!, Fast Fact, What Can I Do?, Connections, Resources

Scientific Inquiry	Unifying Concepts and Processes	Technology
Modeling and explaining	Modeling to explain patterns of change within systems	Building an inertia/tide model

The Bulge on the Other Side of Earth

Background

The bulge of water on the side of Earth that faces the Moon is easily explained. It is due to the gravitational attraction between the Moon and Earth, including the water on Earth. The difference on the horizontal component of gravity results in water moving toward the Moon and creating a "bulge" on the surface of Earth.

The bulge on the other side of Earth is due to **inertia**. Inertia is the tendency of an object at rest to stay at rest and the tendency of a body in motion to continue its motion in a straight line. Pushing a stalled car is extremely difficult because such a massive object has a tremendous amount of inertia, which must be overcome before it will move. Also because of inertia, the car is very difficult to stop once it has begun moving.

Inertia is the reason we wear seat belts in cars. When the car comes to a sudden stop, the people inside tend to keep moving. Objects moving in a circle (or any curved path) tend to keep moving, too. (See **Figure 16.1**.) However, they tend to keep moving in a straight line, not a circle. (The law of inertia was devised to explain this phenomenon.) Some force must be applied to keep the object from going in a straight line. Earth's gravity keeps the Moon moving in an almost circular orbit. Without the gravitational attraction between Earth and the

Vocabulary

Inertia: The tendency of matter to stay at rest or move uniformly along a straight line unless acted upon by an external force.

Fast Fact

The biggest tidal range is observed in the Bay of Fundy in Nova Scotia, Canada, with an average range of 11.7 m. The largest spring tides can reach 16.3 m (recorded in November 1998) at Minas Basin, while the neap tides are "only" 9 m. The second largest observed tidal range was in Leaf Basin, Ungava Bay (in March 2002), with a range of 16.2 m during spring tides. Neither site has ever been measured during extreme highest tides, which will occur next in 2014; the difference is smaller than the observational error (in other words, the difference between the tides at both places is smaller than the error associated with the measurements). Therefore, both the Bay of Fundy and Ungava Bay have the world's largest tides, and need to be considered equal.

Objective

Demonstrate how the rotation of the Earth–Moon system accounts for the bulge of water on the side of Earth facing away from the Moon.

Activity 16

Figure 16.1
Objects moving in a circle tend to move in a straight line. The ball in this diagram, when the string is cut, keeps moving in a straight line.

Vocabulary

Axis of rotation: A straight line about which a body or system of bodies rotates.

Center of gravity: The point in a body or system of bodies at which the entire weight seems to be concentrated.

Figure 16.2
The axis of rotation of the Earth–Moon system is shown as a dashed line, as it passes through the center of gravity of the Earth–Moon system, which is located within Earth.

Moon, the Moon would fly off into space. If a car goes around a curve, the passengers feel they are pushed against the door. Due to their inertia, they tend to keep moving in a straight line, and the door applies a force to make them follow the same curve as the car.

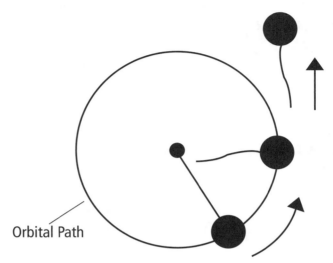

Orbital Path

To see how inertia accounts for the bulge on the side of Earth away from the Moon, it is important to understand that Earth actually undergoes two types of rotation. First, it rotates on its own axis. Second, the Moon and Earth form a two-body system that rotates about a different axis. The **axis of rotation** is located at the **center of gravity** between the Moon and Earth. Since Earth is so much more massive than the Moon, the center of gravity is actually located within Earth. (You might have learned about this in Activity 15.)

The locations of the two axes and the difference between them are shown in **Figure 16.2**. An understanding of this second type of rotation is important in explaining the bulge on the other side of Earth, the side away from the Moon.

Moon

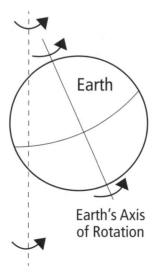

Earth

Earth's Axis of Rotation

172

Procedure

Part 1

1. Put on safety glasses or goggles.

2. Assemble the apparatus as shown in **Figure 16.3**. The vertical dowel (the longer of the two) that runs from the bottle through the 15 cm ball should be inserted 2.5 cm from the surface of the ball, as shown in the diagram. The purpose of the masking tape is to keep the Styrofoam ball from sliding down the stick. Each string should be about 35 cm long. A simple way to attach the strings to the ball is to partially unfold a paper clip and stick the unfolded end into the Styrofoam. (See Procedure 3 of Activity 15.) One end of the string can then be tied to the folded end and the other end to the weights. Heavy-duty paper binder clips work well for the weights.

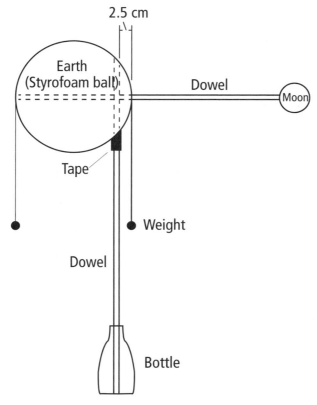

3. Discuss with your group what you think will happen to the strings if you rotate the model. Holding the bottle with one hand, rotate the vertical dowel gently so that the whole apparatus spins. (You may need to practice a little before it works properly. Once you get it spinning smoothly, check with your teacher to make sure it is behaving correctly.)

4. Observe how close the weighted strings come to being horizontal. Record your observations and draw a picture similar to **Figure 16.3**, showing the positions of the strings while the ball is spinning. Which string comes closer to the horizontal?

Figure 16.3
Diagram of the apparatus showing the relative positions and sizes of the different parts (Earth, Moon, axis)

Activity 16

What Can I Do?

Animals that live on the beach exposed by low tides (the intertidal zone) face long odds. Lots of animals forage on the beach—birds, raccoons, people. Do what you can to eat seafood sustainably. For information on best choices, refer to the Monterey Bay Aquarium's Seafood Watch for your part of the world. There is information online and a pocket-sized version to carry with you. Search for "Seafood Watch" at *www.montereybayaquarium.org*.

Part 2

5. Take the strings off the ball and reattach them as shown in **Figure 16.4** (perpendicular to the axis connecting Earth and the Moon).

6. Discuss with your group what you think will happen to the strings now. Spin the dowel again and observe what happens to the strings. Record and draw your observations as before.

Part 3

7. Remove the dowel and reinsert it in the center of the ball as shown in **Figure 16.5**.

8. Discuss with your group what you think will happen now. Spin the dowel and record your observations.

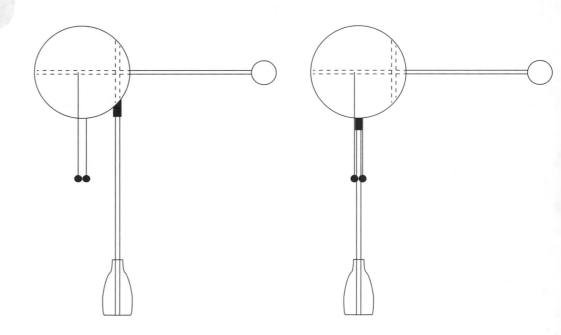

Figure 16.4
Connect the strings to the sides of the Earth ball, perpendicular to the dowel connecting Earth and the Moon.

Figure 16.5 (far right)
Place the vertical dowel in the center of the Earth ball.

The Bulge on the Other Side of Earth

What Is Happening?

The Moon and Earth form a two-body system that rotates about an axis located within Earth. The model constructed in this Activity illustrates the inertial tendency resulting from the rotation of the Earth–Moon system for objects (water among them) to move away from both sides of Earth—the side facing toward the Moon and the side facing away. Finally, the model demonstrates that the effect of things moving away from Earth is much greater on the side facing away from the Moon.

Many textbooks and other sources use the concept of "centrifugal force"—which is actually a misconception—to explain the effects of inertia described in this Activity. According to this misconception, there is a force that acts on all objects that are in circular motion, and this force pushes or pulls the object out from the circle. There is no such force. The misconception arises from our own experience with circular motion. For example, when we go around a curve (part of a circle) in a car, we feel we are being pushed to one side of

How Do We Know This?

How do we predict tides accurately, given the number of variables involved? The problem of predicting the tides remains challenging.

For example, the tidal datum (the reference for tidal elevations) is calculated and changed by NOAA every 19 years, but not at the same time in every location; sandbars change locations; there are parts of the coast that are changing because of the decrease in sediment supply from the continent; and, of course, there are the effects of climate change.

Tidal charts and tidal tables are only available for a few locations in the United States. When we want to know the tidal elevation over a sandbar (not anywhere close to an observing station), we rely on tidal predictions models. These models are as good as the information we feed them. The global tidal forcing is well observed by satellites, but the bathymetric (sea floor) information is still sufficiently deficient to occasionally cause severe problems in the tidal models that can result in boats getting stuck at low tide. The motion of sandbars changes the phase of the tide and the amplitude near the coast. It affects the friction the tides feel and, therefore, the average water level at the coast in the proximity of the changing sandbar.

Objective

Demonstrate how the rotation of the Earth–Moon system accounts for the bulge of water on the side of Earth facing away from the Moon.

Key Concept

II: Ocean structure and water movement

Materials

Each group will need

- safety glasses or goggles (for each student)
- Styrofoam ball (15 cm diameter)
- Styrofoam ball (6 cm diameter)
- string
- dowel (1 m long and 0.5 cm diameter)
- dowel (0.5 m long and 0.5 cm diameter)
- masking tape
- two weights (about 15 g each)
- 473 ml (16 oz.) drink bottle
- ruler

Time

50 minutes

the car. This "feeling" has been incorrectly described as centrifugal force. The sensation is actually due to our bodies' inertial tendency to continue motion in a straight line rather than going around the curve.

The difference between the effects of inertia and the mythical centrifugal force are easily illustrated. Imagine a ball attached to a string that is swinging in a circle. If the string suddenly breaks, centrifugal force predicts what is illustrated in **Figure 16.6a,** which is incorrect. Inertia predicts a path for the ball as illustrated in **Figure 16.6b,** which is correct. See Reading 3: The Tides: A Balance of Forces.

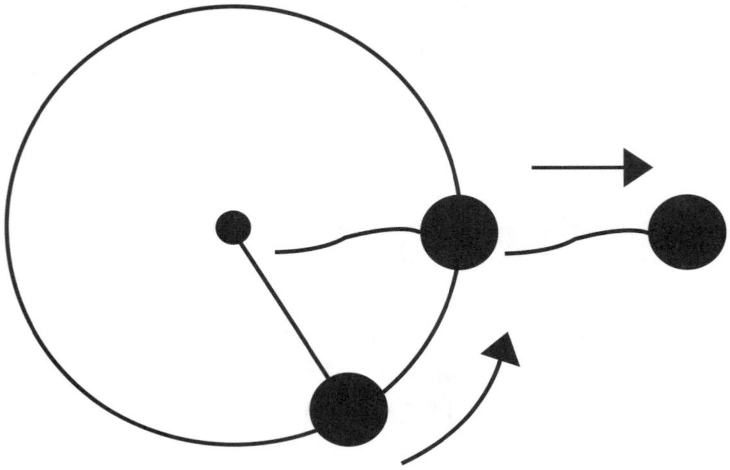

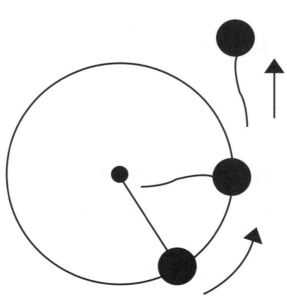

Figure 16.6a
Incorrect trajectory of the ball after the string breaks under the "mythical" centrifugal force

Figure 16.6b
Correct trajectory of the ball after the string breaks under realistic inertial conditions

Preconceptions

This Activity goes hand-in-hand with Activity 15. It adds the concept of inertia, which is sometimes difficult for students to understand. Ask students first to give you an example of inertia, if they can, and the definition of it. Then ask them to relate their example to the definition. The following are some possible preconceptions:

- Tides are caused by the Sun's rotation.
- Centrifugal force is a real force.
- All solar tides are diurnal and all lunar tides are semidiurnal.
- Lakes do not have tides because they are not connected to the ocean.

What Students Need to Understand

- Earth and the Moon form a two-body system that rotates on its own axis, independent of Earth's axis of rotation.
- The bulge on the side of Earth facing away from the Moon is due to inertia.
- Inertia contributes to both tidal bulges on Earth, but much more to the bulge facing away from the Moon.

Preparation and Procedure

SAFETY ALERT
Safety glasses or goggles are required for this Activity.

The purpose of this Activity is to show how the rotation of the Earth–Moon system accounts for the bulge of water on the side of Earth opposite the Moon. The water on the surface of Earth is rotating with Earth just as the strings rotated with the Styrofoam ball. According to the law of inertia, the water tends to move in a straight line away from the surface of Earth rather than in a circle, just as the strings moved away from Earth when the ball was spinning. Without Earth's gravity, the water in the ocean would fly off the surface and into space. If the strings had not been attached to the ball, they too would have flown off. The net effect of gravity and water's tendency to move in a straight line is the creation of a bulge of water on the side of Earth opposite the Moon.

Part 1 of this Activity demonstrates several things. First, the strings move away from "Earth" on both sides, but the string opposite the Moon moves farther away. **Figure 16.7** shows what this might look like. This is because the axis of rotation is not located in the center of "Earth," but is shifted toward the "Moon." The same thing happens with the tides. A bulge due to inertia is created on both sides of Earth, but it is much smaller on the side facing the Moon. This smaller bulge, however, adds to the one created by the gravity of the Moon, making it roughly the same size as the bulge on the other side of Earth.

Part 2 of the Activity, along with Part 3, demonstrate another important aspect of tides. In Part 2, the strings move away from Earth, but they also move away from the Moon, as shown in **Figure 16.8**. This effect is more noticeable when Part 3 is done. In Part 3, the strings move only straight out from Earth, as shown in **Figure 16.9**. The difference is due to the location of the axis of rotation. In Part 2, the axis shifts toward the Moon. Everything on the side of the axis opposite the Moon moves away from the Moon. The same thing happens with the tides. All the water on Earth that is on the side of the axis opposite the Moon tends to move away from the Moon and toward the other side of Earth. This means that the inertial tendency of the majority of water on Earth is to move toward the side away from the Moon, and this explains why such a large bulge can be created there.

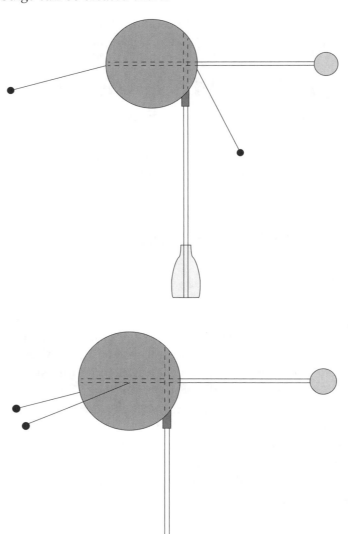

Figure 16.7
Results of Part 1:
Strings move away from
Earth, with the one away
from the Moon being
closer to the horizontal.

Figure 16.8
Results of Part 2:
Strings move away from
Earth but also from
the Moon.

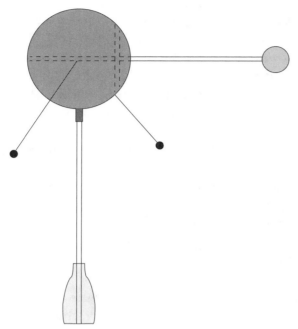

Figure 16.9
Results of Part 3:
Strings move away from
Earth exclusively.

Since Earth has its own rotational axis independent of the rotation of the Earth–Moon system, this Activity can be confusing. A common preconception is that one of the tide-causing factors is Earth's own rotation. Many textbooks explain that tides are caused in part by Earth's rotation, but they fail to specify that it is Earth's rotation in the Earth–Moon system. The rotation of Earth on its own axis does not produce tides. Students and adults alike can have a difficult time distinguishing between these two types of rotation.

You can use two different sizes of sponge balls instead of the Styrofoam balls. The Styrofoam balls can be found in varying sizes at most craft stores. An example apparatus can be set up before the Activity, and you can demonstrate spinning the dowel. Spinning it smoothly requires some practice. You can also do this Activity as a demonstration. Allow ample room for the clips to swing out.

Time Management

Students can do this Activity in 50 minutes, including time for construction and discussion. If they are inclined to ask "what if" questions, be prepared to extend this to another day.

Extended Learning

• If students ask questions or make statements that begin with "I wonder what would happen if we used … [larger balls, heavier balls, a longer distance between Earth and the Moon, etc.]," respond with "How could you set that up? Do you have any insights into what would happen?" Recognize though that they might be asking the question to develop an understanding. If their idea for an experiment is safe, encourage them to try it.

- Have two students stand facing each other, toe to toe. Then have them hold hands and, while keeping their backs and legs straight, lean backward. If students of different weights do this, you can see that one person will be leaning back farther than the other. That is, their center of gravity is not equidistant between them.

- The water on Earth is not the only thing subject to tidal forces. The land is affected as well. However, since land is much more rigid than water, it is far more difficult to deform. In the same way, the Moon is subject to tidal forces. Have students investigate the objects in the solar system that are affected by tidal forces, and learn what those effects are. In some cases, the effects are dramatic. (For example, research the Shoemaker-Levy comet impact. In 1994, the comet broke up because of gravitational forces between it and Jupiter, the Sun, and Jupiter's moons.)

- Encourage students to investigate on their own the difference between "centrifugal force" and inertia. They can do this by constructing a sling that will hold a tennis ball and trying to hit a target with it. It will not take many attempts to convince them that once released, the tennis ball does not fly out from the center of the circle but rather in a straight line, tangent to the circle. (See Reading 3: The Tides: A Balance of Forces.)

- The same factors that operate between Earth and the Moon to create tides also act between Earth and the Sun. Students can investigate how the factors are similar and different in the Earth-Sun system and what impact the factors have on tides on Earth.

Interdisciplinary Study

- Ask students to find examples of authors who use the turning of tides as a metaphor. They can find examples in science, literature, economics, and history. A quick search online will turn up many instances. For example, Brutus, in Shakespeare's *Julius Caesar* says, "There is a tide in the affairs of men."

- Our Moon is moving away from us at about 3.8 cm/year. Geologists had inferred this from growth bands in corals and from pre-Cambrian tidal deposits in the Sangre de Cristo mountains that implied stronger and higher tides in the past. Now, though, we actually can measure our Moon's departure. Ask students to explore the Lunar Laser Ranging Experiment begun by *Apollo 11* astronauts on July 21, 1969.

- Since the Moon is fleeing, ask students to write a science fiction story about what life will be like in the future, say in 100 million years, in the intertidal zone.

- President Kennedy challenged Americans to send a man to the Moon in a joint session of Congress on May 25, 1961. This would require a spacecraft to leave Earth's orbit. Ask students to research the speech and its impact on American life. For photos and an audio recording of the famous speech, go to *www.jfklibrary*.org and search for "man to moon." Students could also learn about the history of space exploration.

Differentiated Learning

- This Activity uses a model to help students learn intuitively why tides are often diurnal. For students who need a comprehensive explanation, the National Oceanic Service provides a podcast about tides and their causes. Search for "NOS podcast Tides" for this 15:33 recording at *http://oceanservice.noaa.gov/podcast/supp_apr09.html#tides*.

- For students who enjoy mathematical challenges, Moon's departure from Earth can provide a good math problem. The Moon's orbital radius increases at 3.8 cm/year. Knowing that, calculate its radius at the time of the dinosaurs' extinction, 65 million years ago. Determine when it separated from Earth (and learn why). Also, discuss assumptions used to make these extrapolations.

Assessment

- Ask students to compare their expectations in Parts 1, 2, and 3 to their observations. Were they surprised, and do they have questions? Are there questions they could explore with the model?

- You can also ask students to summarize for friends what they have learned about the causes of tides; they can present their summary in a paragraph, concept map, video, or podcast. Include descriptions of what they have seen in this Activity (and in Activity15) that has led to their current understanding.

Connections

There are tides on other planets in the solar system. For example, on Jupiter's moon, Io, the internal heating produced by tides is so strong that Io suffers from global-scale extreme volcanism. The combination of the huge mass of Jupiter and the presence of two additional large moons nearby (Europa and Ganymede) creates such a strong body tide. Strong body tides can result in instabilities in the crust and therefore increased volcanic activity. The body tidal bulge is around 100 m (for comparison, Earth's largest body tide bulge is 0.5 m).

Resources

http://www.montereybay aquarium.org

www.jfklibrary.org

http://oceanservice.noaa.gov/ podcast/supp_apr09.html#tides

Activity 17 Summary

Students experiment with a variety of materials and methods for containing and cleaning up an oil spill on water. Questions guide them to evaluate the methods.

Activity	Subject and Content	Objective	Materials
Oily Spills	Oil spills and their mitigation	Explore the effectiveness of different methods for cleaning up oil spills.	Each group of four or more will need: dishpan or plastic tub, tap water, vegetable oil or heavy olive oil (500 ml), cotton string (approximately 1 m long), several drinking straws (cut in half), paper towels, several small pieces of polystyrene foam (such as packing material), 10 ml liquid detergent, sand (125 g), diatomaceous earth (125 g), feather

Time	Vocabulary	Key Concept	Margin Features
50 minutes	Dispersant	III: Impact of human activities on the oceans	Safety Alert!, Fast Fact, What Can I Do?, Connections, Resources

Scientific Inquiry	Unifying Concepts and Processes	Technology	Personal/Social Perspectives	Historical Context
Modeling, experimenting, observing, and analyzing	Modeling to explore systems	Testing technology to clean up spills	Risks and benefits of using oil	Historic oil spills

Oily Spills

Background

Telephones, clothing, skis, antihistamines, ballpoint pens, music cassettes, toilet seats, antifreeze, and gasoline: What do these items have in common? All are often made from oil. The United States, like other industrialized nations, bases much of its economy on numerous products made from oil. This creates a huge demand for oil. To satisfy that desire, natural deposits of oil must be located and extracted. Currently, much of the world's oil comes from areas such as the Gulf of Mexico, Canada, and the Middle East, where natural oil deposits have been discovered and are currently being extracted. The oil must then be transported over large distances in oil tankers or pipelines. When extracting, storing, handling, or transporting oil, there is always the chance of an accidental spill. Oil spills often damage physical environments and harm species.

There are two properties of oil that contribute to its damaging effects on the environment. The first is that, unlike many substances, oil does not dissolve in water very well. The second is the tendency for oil to float on the surface of water. Oil is less dense than water, and less dense materials tend to rise above those that are more dense, like salad oil in oil and vinegar dressing (think back to Activity 6). When spilled into water, large quantities of undissolved, floating oil form slicks that threaten fish, waterfowl, and other animals that live or breathe at the water's surface. Eventually, much of the oil may become denser than the water from interacting with the **dispersants** and sediments, causing it to sink to the bottom.

Vocabulary

Dispersant: A chemical mixture that breaks up oil, keeping damaging slicks from forming and allowing natural processes to degrade oil. Dispersants may be harmful to the environment as well.

Topic: ocean pollution
Go to: *www.scilinks.org*
Code: PESO 012

Objective

Explore the effectiveness of different methods for cleaning up oil spills.

Activity 17

Figure 17.1
String and straws used to construct boom

Some oil may also form "tar balls" that wash ashore. Tar balls and slicks that wash up on land impact shorelines, destroy plant and animal life, devastate wildlife habitats, and adversely affect human activities and industries.

There are several methods available for cleaning up oil spills, either breaking down (degrading) the oil, or limiting environmental damage caused by the oil. Floating booms are used to contain the oil. Detergents and other chemicals—known as dispersants—are used to act against the oil and break it up. The effectiveness of any method depends largely on the type of oil spilled, the size and location of the spill, and the prevailing weather conditions in the area. Other circumstances, unique to each spill, also affect the selection of a cleanup method or combination of methods.

In this Activity, you will simulate an oil spill in freshwater, and experiment with various methods of oil spill cleanup.

Procedure

1. Fill the dishpan or sink half full with tap water.

2. Construct a boom using the straws and string. (See **Figure 17.1**.) A boom is used to create a border around the spill in order to contain it.

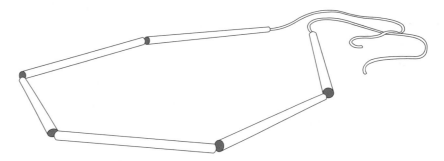

3. Pour a small amount (slightly more than 100 ml) of oil into the water to simulate an oil spill. Try to contain or remove the oil using the boom. Note how a slick is created. (If necessary, replenish the water and oil in the pan or sink before trying each new method. Try to use nearly the same amount of oil each time.) Record your comments about the effectiveness of this method in the data table on **BLM 17.1**. (Use this data table to record information in each step of this procedure.)

4. Use a paper towel to gently blot or skim the oil from the surface of the water. Record your comments about the effectiveness of this method in the data table.

5. Place the pieces of polystyrene foam on the oil slick. After a minute or two, remove the polystyrene foam. Record what happened to the slick in the data table.

6. Add a drop of detergent to the surface of the oil near the center of the slick. Note what happens to the slick and record your observations in the data table.

7. Sprinkle some sand on the entire surface of the spill and note what happens. Record your comments in the data table.

8. Sprinkle some diatomaceous earth over the surface of the spill and observe what happens. Record your comments in the data table.

9. Devise one or more additional methods for cleaning up the oil. Have your teacher review your methods, then try them out and record what happens in the data table.

10. In the data table, rank the effectiveness of each cleanup method by numbering the methods from 1 (most effective) to 5 (least effective).

11. Dip the feather into the oil and note what happens to it. Record your observations in the data table.

12. Try rinsing the oil from the feather using tap water alone, then using tap water with soap. Record your observations in the data table.

13. When you have finished your investigation, clean the work area and discard used materials.

Questions and Conclusions

1. Which cleanup method was the most effective in removing the oil spill?

2. What problems might be encountered when using this method to clean up a real oil spill that has spread over a large area?

3. How might weather conditions affect the usefulness of this method?

4. Describe the problems you encountered using each of the other methods.

5. What factors (besides the size and location of the spill and the weather conditions) might affect the success of an oil spill cleanup operation?

6. Based on your observations of the feather after it was dipped into the oil, what do you think would happen to birds that were coated with oil as a result of a spill?

7. No attempt at controlling an oil spill is likely to be 100% effective. For example, in 1989, the oil tanker *Exxon Valdez* ran aground on a reef in Prince William Sound off the coast of Alaska, spilling more than 10 million gallons of oil into the water. It is estimated, for example, that less than one-half of the oil from the *Exxon Valdez* oil spill was recovered.

    Natural processes will slowly break down the oil that remains in the environment. It is possible to speed up the removal of oil from the environment using methods such as beach cleaning and the addition of chemicals to the environment. These techniques are costly and time-consuming. Do you think the investment of time and money should be made to speed up the breakdown of oil in the environment? Why or why not?

8. What do you think are the causes of most oil spills? Can you think of some ways in which oil spills could be prevented?

Fast Fact

The Deepwater Horizon spill in the Gulf of Mexico during the spring of 2010 was the largest in U.S. history, spilling over 200 million gallons over a period of 87 days. Although called a "spill," it was actually caused by an explosion of the drilling platform that resulted in a leak at the sea floor. The leak was especially difficult to stop because the well was drilled in a location where the ocean was approximately 5,000 ft. deep, and repairs had to be made underwater at that depth. Cleanup efforts and work to determine the extent of impacts will span at least a decade.

What Can I Do?

If you live along the coastline, walk the beaches at least once a month and look for tar balls in the sand, or wildlife (generally shore birds) that appear to have been soaked with oil. If you see these, contact your city or county environmental agency. Even if you do not live along the coast, you can strive to minimize your use of petroleum products.

Data Table: Oil Spill Cleanup Methods

Cleanup Method	Comments on Effectiveness	Method Rating

Oily Spills

What Is Happening?

The effectiveness of controlling and cleaning up oil spills depends largely on the type of oil spilled, the size and location of the spill, and the prevailing weather conditions in the area. In this Activity, students will simulate an oil spill in freshwater, and then evaluate various methods of oil spill cleanup. Remind students to draw on their knowledge from Activities 1, 2, 3, and 5 from this textbook, which deal with the chemistry and structure of water, to understand the relationship between oil and water.

Each of the cleanup methods students will use has a real-life counterpart. The straws and string used to make booms to contain the spill are very similar to the real-life booms used to surround oil spills for containment. Paper towels model absorbent cloth, which is used to soak up oil that has washed ashore. Polystyrene foam, sand, and diatomaceous earth all act to soak or attract the oil. Detergents act to disperse the oil so slicks do not form.

This Activity also provides an opportunity for discussing how oil that remains in the environment should be eliminated. Natural processes will eventually degrade the oil, but the processes occur very slowly, over years or decades. Physical cleaning can remove some of the oil from rocks and beaches. For instance, oil that reached shore from the Deepwater Horizon and *Exxon Valdez* spills was often cleaned by hand with absorbent cloth. Chemical dispersants can speed up the degradation of oil, as was especially obvious during the Deepwater Horizon incident. Microorganisms that consume oil are now used to combat some spills. Several of these cleanup techniques can be used on the same spill, but they are all costly and time-consuming. Debates arise as to whether such large investments into cleanup should be made and who is responsible for providing them.

How Do We Know This?

How can we determine if oil in the ocean originated from a natural source?

Oil in the ocean may originate from a natural source, called a seep, or from an unnatural source, such as a spill. Since all oil has a unique geochemical "fingerprint" representing its source, and we can detect this "fingerprint" using geochemical techniques, it often is possible to determine the specific source of oil, and in doing so establish whether an occurrence of oil was natural or caused by human activity.

Objective

Explore the effectiveness of different methods for cleaning up oil spills.

Key Concept

III: Impact of human activities on the oceans

Materials

Students will work in groups of four or more for this Activity.

Each group of four or more will need

- dishpan or plastic tub
- tap water
- vegetable oil or heavy olive oil (500 ml)
- cotton string (approximately 1 m long)
- several drinking straws (cut in half)
- paper towels
- several small pieces of polystyrene foam (such as packing material)
- 10 ml liquid detergent
- sand (125 g)
- diatomaceous earth (125 g)
- feather

Time

50 minutes

Preconceptions

If students did Activities 2 and 5, ask them what they know about oil and water—in salad dressing, for instance. Then ask them to remind you of what they learned in those two Activities. (The two Activities refer to the polarity of water molecules, and oil is a nonpolar substance. Chemists say that "like dissolves like": polar solvents dissolve polar substances, including ionic compounds such as salt. Polar solvents do not dissolve nonpolar substances such as oil very well.) If students did not do those Activities, ask them, "From your everyday experience, how well do oil and water mix? Give an example." Students might have the following preconceptions:

- When you dump stuff in the ocean, it just disappears.
- Improved designs in oil tankers keep them from leaking oil.
- Only oil tankers are at risk of leaking oil.
- Oil spilled on land cannot affect the ocean.

What Students Need to Understand

- For the most part, oil and water do not mix because water is a polar molecule and oil is nonpolar. Nonpolar substances will not dissolve readily in water.
- Nearly all types of oil float on water because they are less dense than water.
- Many factors affect the selection of an oil cleanup method and the outcome of the cleanup operation.
- Oil in the environment after a spill may be left for natural processes to degrade, or it sometimes can be removed using costly cleanup methods.
- Oil spill cleanup is a complex, time-consuming, and messy activity.

Time Management

This Activity can be completed in 50 minutes. Be sure to leave 5 to 10 minutes for students to clean up and for you to check that their dishpans or sinks have no oily residue that will spoil future labs.

Preparation and Procedure

Prior to beginning the Activity, collect and distribute the materials. If some of the materials listed are not available, the following items may be substituted: sponge or grass straw (both for absorbing the oil); aquarium net (for skimming). Alternatively, students may bring their own materials to try as oil removers.

SAFETY ALERT

1. Diatomaceous earth can be an extremely hazardous material with potentially serious health hazards. Teachers are encouraged to use only food grade diatomaceous earth or alternative such as ZeoFiber or AquaPerl.

2. Be careful to quickly wipe up any spilled liquids on the floor—slip and fall hazard.

3. Wash hands with soap and water upon completing the lab.

4. Review with students pertinent safety procedures and required precautions for hazardous materials found in the MSDSs.

Extended Learning

- Simulate spill cleanup under varying weather conditions: add ice to the water to simulate an Arctic spill; modify the water temperature to illustrate differences between spills in temperate and tropical locations; use a portable hair dryer to illustrate complications of cleanup caused by wind.

- Teachers' Domain has a simulation of an oil spill, which includes the impact of wind. Search for "Teachers' Domain oil spill simulation" at *www.teachers domain.org/ext/ess05_int_oilspill/index.html.*

- Use salt water to simulate ocean water, and compare cleanup methods for freshwater and ocean waters. Investigate actual oil spills around the world and the methods used to contain them.

SAFETY ALERT

Use care when using electricity near water. The use of a portable stand on which to support the hair dryer is recommended.

Interdisciplinary Learning

- Explore the economic effects on communities near oil spills. (Examples include the 2010 Deepwater Horizon spill, the 1989 *Exxon Valdez* spill, a 1969 spill near Santa Barbara, California, and the 1991 release of oil into the Persian Gulf during the Persian Gulf War.) In addition, investigate the impact that oil spills have on sea life and coastal communities, and debate who is responsible for cleaning up oil spills and who will pay for the price of cleanup.

- Have students explore the international politics and economics of petroleum. They can research which countries produce oil, which countries export it, and which countries import it. They can also investigate how the petroleum is used. Search for "oil producers," "oil exporters," and "oil importers." Suggest to students that they evaluate the reliability of the websites that the search engine finds.

- Ask students to research the use of microorganisms in mitigating spilled oil, "bioremediation." These are new techniques that scientists and engineers are developing. Have your students search for "bacteria oil spills," "microorganism oil spills," and "bioremediation" online and in any electronic databases that your media center has. See *www.nature.com/news/2010/100916/full/ news.2010.475.html, www.scientificamerican.com/article.cfm?id=how-microbes-clean-up-oil-spill*s, and *http://response.restoration.noaa.gov.*

Connections

- The La Brea Tar Pits in southern California are a natural seep of petroleum in which Pleistocene animals were trapped, died, and fossilized. Ask students to learn about the geology and paleontology of this fascinating window into the Ice Ages at Rancho La Brea. Search for "La Brea Tar Pits" at *www.tarpits.org.*

- Bioremediation is a safe and effective way to clean up oil spills. Search for "Bioremediation" at *http://response.restoration. noaa.gov.*

Differentiated Learning

- After students do this engaging hands-on Activity, many will benefit from seeing the effects of real spills. Teachers' Domain offers a 4-minute-and-44-second segment from NOVA's program, "The Big Spill." The segment focuses on the *Exxon Valdez* spill, but also shows lingering effects from a spill on Cape Cod. Captioning is an option. Search for "Teachers Domain Oil Spill: Exxon Valdez, 1989" at *www.teachersdomain.org/resource/ess05.sci.ess.watcyc.exxon.*

- You can challenge students who love mathematics to do an experiment that would allow them to extrapolate the area of a spill created by a quart of oil or a 55-gallon drum of oil. This project would involve designing an experiment with multiple trials and the use of a graph. Students could start by determining the number of drops of oil in a milliliter and the area produced by one drop.
- Students can write a persuasive piece, agreeing or disagreeing with a writing prompt such as, "Oil drilling in an environmentally sensitive area is essential to maintaining our standard of living." Examples of environmentally sensitive areas are the Arctic National Wildlife Refuge and continental shelves. See *http://arctic.fws.gov*. You might choose to collaborate with a language arts teacher on this writing assignment.

Answers to Student Questions

1. Answers may vary. If different groups found different methods most effective, discuss possible reasons for the differences.

2. Answers will vary. For the boom method: it is difficult to create a continuous boom around a very large area. For absorption methods: covering the entire area may prove difficult. For dispersion methods: having a sufficient amount of dispersant may be a problem, and some dispersants may be harmful to fish and waterfowl. Also, dispersants are often applied by airplanes and can be deflected away from the spill by the wind as they are applied. All methods are affected by currents, obstructions (such as islands, shorelines, docks, buoys), and shallow water where cleanup vessels are unable to navigate.

3. All methods will be affected by extremely high or low temperatures, rough seas, high winds, storms, lightning, rain, hail, tides, and strong currents.

4. Answers will vary depending on observations.

5. Other factors affecting the success of an oil spill cleanup are how soon after the spill occurs that cleanup begins; the availability of the resources (people, money, and equipment) necessary to provide a continuous cleanup effort; the experience and training of the people involved; whether oil continues to spill after cleanup begins; how close to shore the spill occurs; the geography and geology of the area where the spill occurs (water temperature, an inlet versus open water, air temperature, the topography of the bottom, the depth of the oil source).

6. Answers may vary. Birds whose feathers become coated with oil have difficulty flying, swimming, and gathering food. They may drown, be preyed

upon, or starve to death if the oil is not removed. Also, their feathers lose much of their insulating properties and the birds may lose so much heat that they die of exposure to cold. They may also poison themselves as they try to clean their feathers and end up ingesting the oil.

7. Answers will vary. This question provides opportunity for discussion of the advantages and disadvantages of removing spilled oil from the environment.

8. Answers will vary. Some causes of oil spills are tanker accidents (running aground, colliding with another vessel, splitting the hull); pumping station accidents; leaks in storage tanks; and pipeline leaks. In coastal environments, small recreational boating spills are often a major source of spilled oil. Prevention of oil spills involves, among other things, special tanker construction to guard against leaks in the event of an accident; training of tanker, pump station, and pipeline personnel in the proper methods of handling oil-related equipment; and proper maintenance of oil-handling vessels and equipment.

Assessment

- For formative assessment, circulate among students as they do this Activity. Listen for comments within their groups for potential spin-off trials, and follow up on questions they ask (e.g., "Would it work better if we used…?").

- For summative assessment, you can grade the answers to the above questions. Alternatively, you could assign the writing prompt in Differentiated Learning as an essay question. You could ask students to write a fictitious news report—an article, blog, or script for a recording—about an oil spill. This could include the accident or cause of the spill, its extent, its immediate consequence, methods of cleanup being used, and anticipated long-term consequences.

Resources

www.teachersdomain.org/ext/ess05_int_oilspill/index.html

www.nature.com/news/2010/100916/full/news.2010.475.html

www.scientificamerican.com/article.cfm?id=how-microbes-clean-up-oil-spills

response.restoration.noaa.gov

www.teachersdomain.org/resource/ess05.sci.ess.watcyc.exxon

http://arctic.fws.gov

http://www.tarpits.org

http://response.restoration.noaa.gov

Activity 18 Summary

Students experiment with a variety of materials in water and soil to compare their durability and, conversely, their degradability. They investigate which float in water and which break down in water and/or soil.

Activity	Subject and Content	Objective	Materials
Forever Trash	Decomposition of ocean pollutants	Observe the breakdown of various materials in water and in sand.	Each group will need: small piece of paper (10 cm × 10 cm); 10 cm × 10 cm scraps of cloth (cotton, rayon, wool, polyester, nylon, etc.); small sheets of aluminum foil, waxed paper, plastic wrap; aluminum soda can tabs; plastic bag (sandwich size); pieces of a plastic grocery bag; plastic six-pack holder; plastic bottle cap; hard candy in a plastic wrapper; unwrapped hard candy; rubber balloon; polystyrene foam packing peanuts; starch-based packing peanuts; sand containing organic matter; shoe box or small individual containers; beaker; salt (for salt water)

Time	Vocabulary	Key Concept	Margin Features
100 minutes (50 minutes twice)	Biodegradable	III: Impact of human activities on the oceans	Safety Alert!, Fast Fact, What Can I Do?, Connections, Resources

Scientific Inquiry	Unifying Concepts and Processes	Personal/Social Perspectives
Experimenting, observing, and analyzing	Experimenting with conditions to explore changes in a system	Risks of designing synthetic materials

Forever Trash   Activity

Background

Stories and legends speak of bodies and treasures buried at sea. Hundreds of thousands of boats and ships and the materials and supplies on them have sunk to watery graves since humans first sailed the oceans. Even today, it is common practice for humans to throw their waste into the seas. In the past, much of that material would degrade and/or decompose in the ocean waters. The composition of waste, however, has changed from natural, **biodegradable** materials to synthetic materials that resist decay.

Use of products that do not decompose has increased dramatically over the past 20 years. As items made of plastic or with plastic packaging have become more and more common, waste disposal has become increasingly difficult to control and regulate. Much of that waste ends up in the oceans. The impact of synthetic debris on the marine environment has become painfully evident. Floating trash is visible in most bodies of water. Bits of plastic, polystyrene foam, aluminum beverage cans, and broken glass can be found on beaches throughout the world.

A series of laws passed by the U.S. government have outlawed dumping of medical waste, sewage sludge—a by-product of sewage treatment—and plastics in the oceans and waters around the United States. International treaties also have tried to set the same standards for all countries, but ocean dumping and pollution remain a problem.

Plastics and other synthetic pollutants are made to last and do not break down readily. This "forever trash" floats in the surface waters, entangling and killing animals, or sinks to the ocean depths, altering the marine environment.

Vocabulary

Biodegradable: Material that can be broken down into simpler substances (elements and compounds) by bacteria or other decomposers.

Fast Fact

On the third Saturday in September, from 1999 to 2004, volunteers around the world picked up a total of 8.2 million cigarettes or filters, 3 million food wrappers, 1.9 million plastic bottles, 1.7 million bags, and 1.6 million glass bottles. This is part of the annual International Coastal Cleanup.

Objective

Observe the breakdown of various materials in water and in sand.

Topic: ocean pollution
Go to: *www.scilinks.org*
Code: PESO 012

Activity 18

Materials

Each group will need

- small piece of paper (10 cm × 10 cm)
- 10 cm × 10 cm scraps of cloth (cotton, rayon, wool, polyester, nylon, etc.)
- small sheets of aluminum foil, waxed paper, plastic wrap
- aluminum soda can tabs
- plastic bag (sandwich size)
- pieces of a plastic grocery bag
- plastic six-pack holder
- plastic bottle cap
- hard candy in a plastic wrapper
- unwrapped hard candy
- rubber balloon
- polystyrene foam packing peanuts
- starch-based packing peanuts
- sand containing organic matter
- shoe box or small individual containers
- beaker
- salt (for salt water)

Time

100 minutes

Figure 18.1

Place each item in a saltwater aquarium and bury an identical item in a sand-filled shoe box.

SAFETY ALERT

1. Wash hands with soap and water upon completing the lab.

2. Never eat food used in the lab Activity or bring any other food or drink into the lab.

Procedure

1. Select a designated number of items to "test."

2. Subject each item to a "sink or swim test." Record your predictions as to whether each item will float or sink in water in the data table on **BLM 18.1**. Test your predictions by dropping the item to be tested in a beaker of water. Record your results in the data table on **BLM 18.2**.

3. Use the following instructions and test identical items for their ability to decay in water and in sand. (See **Figure 18.1**.) Again, record your predictions as to the outcome of the tests in the data table on **BLM 18.1**.

   (a) Place one of each item in a class aquarium filled with salt water. Leave the items in the water for one to two weeks.

   (b) Using individual containers or a "common grave" (a shoe box filled with sand), bury one of each item in sand containing organic material. Leave the items in the sand for one to two weeks.

4. After the allotted time, use a net to retrieve items from the water and a spoon to uncover the buried items. Observe and record the changes in each item in the data table on **BLM 18.2**.

Questions and Conclusions

1. Which items floated?

2. List where floating trash might end up.

3. How can floating trash be harmful to marine life?

4. Which items sank?

5. How might sunken trash be a hazard to marine life?

6. List the items that broke down or decayed (list as "no decay," "some decay," or "heavily decayed") in the water decay test. Explain why these materials behaved as they did.

7. List the items that broke down or decayed most readily (list as "no decay," "some decay," or "heavily decayed") in the sand burial decay test. Explain why these materials behaved as they did.

8. Suggest several ways to limit the impact of pollution on our oceans.

9. How well did your predictions match the actual results you achieved? Explain the differences you observed.

What Can I Do?

If you live along the coastline, volunteer for a beach cleanup effort. Most coastal municipalities sponsor an annual or semi-annual beach cleanup. Contact your city or county environmental agency for details.

Data Table: Predictions

	Item	Sink/Float	Description of Decay	
			in Water	in Sand
#1				
#2				
#3				
#4				
#5				

Data Table: Results

	Item	Sink/Float	Description of Decay	
			in Water	in Sand
#1				
#2				
#3				
#4				
#5				

Forever Trash

What Is Happening?

Trash and debris pose a serious threat to marine wildlife, navigation, and general water quality. Merchant ships, the world navies, commercial fishermen, recreational fishermen, boaters, and beachgoers all contribute to this problem. Human trash, in the form of food, plastics, beverage cans, polystyrene foam, cigarette butts, glass, and so on, is a visible characteristic of marine environments.

Scientists have identified plastic as the single most dangerous threat from humans facing many animals. Production of plastic in various forms has increased dramatically in the past 20 years. The durability of plastic makes it extremely difficult to dispose of. As a result, plastic has become an important marine pollutant. Plastics are often ingested by marine animals or serve as death traps—entangling, strangling, and suffocating the animals.

This Activity explores various types of marine debris. Students will predict and discover which items float or sink, and will observe the rates of decay of several common marine pollutants. Remind students that people are taking steps to address the problem of marine pollution. Laws and treaties (discussed in Reading 5: The Ocean: A Global View) are good examples. Recycling programs for plastics, metal, paper, and other materials are also helping to reduce debris. Reiterate to students that they, too, can play a role in addressing the problem through thoughtful purchasing, and understanding the science behind the problem of marine debris and solutions to it.

Objective
Observe the breakdown of various materials in water and in sand.

Key Concept
III: Impact of human activities on the oceans

Materials
Each group will need
- small piece of paper (10 cm × 10 cm)
- 10 cm × 10 cm scraps of cloth (cotton, rayon, wool, polyester, nylon, etc.)
- small sheets of aluminum foil, waxed paper, plastic wrap
- aluminum soda can tabs
- plastic bag (sandwich size)
- pieces of a plastic grocery bag (There are different kinds—some claim to degrade in light or landfills. Try to find an example of the different types.)
- plastic six-pack holder
- plastic bottle cap
- hard candy in a plastic wrapper
- unwrapped hard candy
- rubber balloon
- polystyrene foam packing peanuts
- starch-based packing peanuts
- sand containing organic matter
- shoe box or small individual containers
- beaker
- salt (for salt water)

Time
100 minutes

How Do We Know This?

Why is plastic trash so dangerous in the ocean?

Plastic has been identified as the single most dangerous trash item to the marine environment. This is not only because it decomposes so slowly, but because when eaten by marine organisms, even small particles of plastic cannot be digested or passed through their digestive system. What eventually happens is that the system becomes blocked and the organism dies of starvation.

Preconceptions

Ask students to list examples of trash they have seen in waterways or at the beach. Then ask them to list what they have heard about permanence or biodegradability of trash. For instance, they might have seen plastic water bottles in ditches and heard that a plastic bottle will stay whole forever. The following are some possible preconceptions students might have:

- The oceans are so large that dumping trash in them will have no effect.
- Garbage will sink to the bottom of the ocean and be out of the food chain.
- When people throw things in waterways or on land, the things will stay where they are.

What Students Need to Understand

- Rates of decomposition are different for different materials.
- Plastics are synthetic materials designed to last. They do not break down readily.
- "Degradable" plastics do not necessarily break down completely.
- Plastics, regardless of breakdown properties, remain a serious threat to marine wildlife.

Time Management

The setup and wrap-up of this Activity will take 100 minutes. The setup can be completed in 50 minutes, with the same amount of time required for the wrap-up session. This includes time for students to clean up and to answer questions.

At least one to two weeks is required between sessions—and it may actually take six to nine weeks for noticeable decomposition to occur.

Preparation and Procedure

Obtain a large quantity of sand in which to bury waste materials. This sand should contain organic matter. If only commercial clean "sandbox" sand is available in your area, leave the sand outside in a bucket or large shallow pan for several weeks or add to it a small amount of organic material, such as pond water or compost. The amount of organic material in the sand will greatly influence the rate of decay in this experiment.

A large receptacle such as an aquarium may be used to model the decay of floating or sunken trash in the ocean. The addition of dissolved salts is suggested in modeling an ocean environment. (Note: This experiment may develop an unpleasant odor, depending on the presence of airborne particles and the composition of the waste materials being tested. A one-week testing period may be too long for some noses.) Know the source for any materials used in this Activity. Never use garbage or other refuse, given the potential for harmful biologicals. Also, check for mold or other fungi growth. Some students are allergic to these biologicals.

Extended Learning

- Some students should add an experimental control by testing material strength prior to decay. Some materials, however, may be too strong to break before they decompose in water or sand.

- Students can do quantitative tests of decay by using equal-sized sheets of cloth, paper, plastic wrap, and aluminum foil. Simply bury each material in sand or water for the same time, retrieve the material, and test its strength. To do this, have two students pick up the material to be tested by all four corners. A third student should place a single penny in the center of the test material and continue adding pennies one by one until the material breaks, tears, or disintegrates. Record the number of pennies necessary to "break" the material. Repeat this procedure for each material being tested. (Materials suggested for quantitative studies include different brands of paper towel, plastic wrap, aluminum foil, waxed paper, newspaper, notebook paper, and cotton, rayon, wool, or nylon cloth.)

- Have students investigate the Great Pacific Garbage Patch. What is the size of it? What are the main components of the Patch? How much trash are we talking about? What are the consequences for marine life? Scripps Institute of Oceanography dedicated a research expedition to study the science of the garbage patch. Search for "SEAPLEX" to learn the purpose and results of this 2009 mission at *http://sio.ucsd.edu/Expeditions/Seaplex*.

- Ask students to compile a list of products that are marketed as biodegradable. They can discuss whether they agree that the marketing is reasonable for the specific products, based on their research in the lab and the research of other scientists.

- Many states have an annual International Coastal Cleanup day, usually during the fall. Coordinated by the Center for Marine Conservation and other state and private agencies, the event focuses on cleaning up trash from beaches and other waterways. Contact appropriate agencies in your state to organize or participate in the cleanup or similar activities. You may want to plan a cleanup near your school or in your local community. (See **Figure 18.2**.) As part of a national citizen science project to monitor estuaries, the Environmental Protection Agency publishes a protocol to measure the kind and abundance of marine debris— trash—on beaches. This is part of the National Volunteer Estuary Monitoring program. Search for "Volunteer Estuary marine debris" at *www.epa.gov/owow/ estuaries/monitor/pdf/chap16.pdf*.

Connections

Ditches, streams, and rivers connect with oceans even when the oceans are thousands of miles away. Debris that begins in a ditch can end up in the ocean and then back on a beach or in the gut of a marine mammal. Ask students to trace the potential path of a piece of trash from a nearby ditch or stream, downstream to the closest salt water. This path connects the studies of hydrologists or surface-process geologists to that of oceanographers and marine biologists. Search for "marine debris" at *www. noaa.gov*.

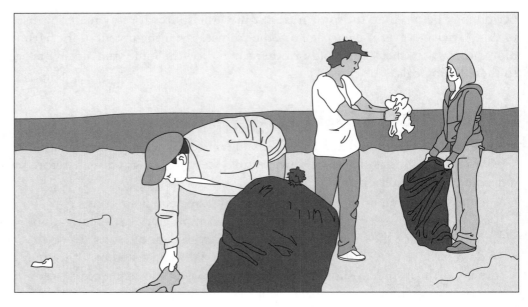

Figure 18.2
Plan a cleanup program in your local community to pick up trash from beaches or along other bodies of water.

Interdisciplinary Study

For years, the sea has held mystery and fascination for humans. Observe people who visit the ocean and you will see that they spend a great part of their time looking seaward, searching. Are they looking at incoming waves to check for swimming conditions, seeking out a speck on the horizon that will later materialize as a ship, or searching for treasures washed up by the waves? Or, do people become spellbound by the soothing sight and sounds of ocean waves repeatedly washing upon the beach? While experiencing the tranquility of watching the Sun rise or set across the ocean waves, do they become lost in the sheer mystery and vastness of the blue horizon? The sea has become a focus for expression in art, music, and literature.

- To provide students with an opportunity to experience the beauty and value of the ocean, and come to understand the importance of keeping the ocean and shorelines free of "forever trash," have students use sandpaper, crayons, and an iron to create a "sea print." The following are instructions for creating a print:

  1. Use crayons to create a sea picture on a sheet of sandpaper, firmly pressing on the crayons to get a thick coat. After the scene is completed, fill in the background with an appropriate color.

  2. Place the colored sandpaper face up on a thick stack of newspapers.

  3. Place a sheet of manila paper over the print. Center the paper so that the print is in the middle.

  4. Slowly press the print with a warm iron until you see evidence of the crayons being transferred to the manila paper.

  5. Trim and frame the print with a contrasting piece of construction paper.

• You could have students learn about careers in modern plastics. For a video clip (2:13) about designing more biodegradable plastics from WGBH carried by Teachers' Domain, search for "Teachers careers plastics" at *www. teachersdomain.org/resource/ate10.sci.engin.systems.plasticsjobs.*

Differentiated Learning

To make this quantitative for students who enjoy mathematics, or to demonstrate to weaker math students why math is useful, you can give students a project in marine debris using real data, with a focus on "ghost" crab pots. These are crab pots that break loose from their buoys and continue to trap fish. Search for "Ghostbusting in the Chesapeake" from BRIDGE, a marine science educational resource at the Virginia Institute of Marine Sciences at *http://www2.vims.edu/ bridge/DATA.cfm?Bridge_Location= archive1010.html.*

Answers to Student Questions

1. Answers will vary depending on items selected. Some items float initially and then become saturated and sink.

2. Floating trash can wash up on beaches and become entangled in boat propellers, trapped among aquatic plants, or ingested by marine wildlife.

3. Plastics and other floating trash items are often ingested by marine animals. They have no nutritional value and interfere with the digestive process. Floating debris may also trap, strangle, or injure marine species.

4. Answers will vary depending on items selected. Some items float initially and then become saturated and sink. The size and the orientation of the material (for example, crumpled versus flat aluminum) may also affect whether a material "sinks or swims."

5. Sunken trash may be ingested by marine life, interfering with the digestive process. Sharp edges, rough surfaces, and narrow openings may harm or injure wildlife.

6. Answers will vary depending on items selected and the length of the test period. Natural products decay much more readily than synthetics. Again, size and the orientation will affect the rate of decay.

7. Answers will vary depending on items selected and the length of the test period. Natural products decay much more readily than synthetics. Again, size and the orientation will affect the rate of decay.

8. Answers will vary. Some ways to limit the impact of pollution on our ocean include use of recyclable and reusable materials, use of truly biodegradable materials, preprocessing waste before dumping, establishing and enforcing government restrictions on ocean dumping, and using alternative disposal methods (landfills, etc.).

9. Answers will vary.

Assessment

- For formative assessment, circulate among students to troubleshoot and to listen for ideas they have about other materials to test or other conditions under which to test them. If their ideas are feasible and safe, encourage them to try out the ideas.

- For summative assessment, you can ask students to tell you what they have learned about "forever trash." Has their awareness of marine debris changed? Have their behaviors changed? If so, in what ways have they changed? This would work well as a journal entry or a Think-Pair-Share.

- Alternatively, you can grade students' answers to the questions.

Resources

www.epa.gov/owow/estuaries/
monitor/pdf/chap16.pdf

http://sio.ucsd.edu/Expeditions/
Seaplex

www.teachersdomain.org/
resource/ate10.sci.engin.
systems.plasticsjobs

http://www2.vims.edu/
bridge/DATA.cfm?Bridge_
Location=archive1010.html

www.noaa.gov

Introduction

The following Readings elaborate on the concepts presented in the Activities: Water: The Sum of Its Parts; The Ocean; The Tides: A Balance of Forces; Waves; and The Ocean: A Global View. Although the Readings were written especially for this volume with the teacher in mind, students should also be encouraged to read them for both interest and additional study.

The ocean plays a fundamental role in many ways: from affecting the global climate and its variability, to phenomena such as El Niño, to the dispersal of pollutants such as oil spills. Understanding the dynamics controlling the ocean is essential to the prevention of potential negative long-lasting effects.

Water: The Sum of Its Parts

People commonly think of the oceans of the world as distinct bodies of water in different regions of the globe. In reality, they are all part of one great world ocean. The concept of a world ocean is important to understanding the world ecology and the effects that activities in one region have on environments and populations elsewhere. In keeping with this view, and to reinforce the notion of one world ocean, the term *ocean* as used in these Readings generally refers to all of the world's oceans.

The beauty and mystery of the ocean have long fascinated children and adults alike. Studies of the ocean have revealed its usefulness, complexity, and importance to Earth's economic and ecological systems. The ocean supplies food, water, chemicals (such as iodine, bromine, and magnesium), fuels, recreational opportunities, and transportation for 90% of the world's international trade. In the United States, coastal regions make up only 15% of land area, but more than half of the human population lives in coastal countries. The five largest cities in the world are located on coastlines. Many countries' communication and defense systems are largely ocean based. These facts suggest a worldwide economic dependence on the ocean.

Perhaps less obvious than its economic significance are the effects the ocean has on Earth's ecosystem. The ocean occupies over 70% of Earth's surface and represents a rich and diverse biome. There is geologic evidence to suggest that life originated and developed in the ocean long before appearing on land. Microscopic organisms that thrive in the ocean supply more life-sustaining oxygen to Earth's atmosphere than any other source, including plants on land. The ocean's role in Earth's climate and weather systems is critical. It serves as a principal source of water for precipitation. It absorbs radiant energy from the Sun, thereby moderating Earth's climate, especially in coastal regions. Ocean currents, along with wind, distribute much of this absorbed heat throughout the globe, thereby affecting the climate in all regions of the world, even those remote from any coastline.

Henry Bryant Bigelow, the first director of the Woods Hole Oceanographic Institution, once said, "The most important thing about the ocean is that it is full of water." Seawater does indeed contain 97% pure water, and this statement calls attention to the crucial role the properties of water play in determining the properties of the ocean.

Water is a unique substance with properties much different from those of other compounds. It is the most common substance on Earth, yet it is the only substance that occurs naturally in all three states, or phases—solid, liquid, and gas. It is also unique in that its solid form is less dense than its liquid form—ice floats in water. Water has a high ability to store heat and dissolves more substances and in greater quantities than any other liquid.

Reading 1

Topic: water
Go to: *www.scilinks.org*
Code: PESO 003

These properties have a tremendous impact on Earth and account for many of the life-sustaining attributes of the planet.

The water molecule is a fairly simple molecule made up of two elements—hydrogen and oxygen. Two hydrogen atoms (H) bond to one oxygen atom (O), creating a single water molecule (H_2O). To understand water's structure and why it exhibits certain characteristics, we must first examine an atom of each of its two elements. A hydrogen atom consists of one proton (in the nucleus) and one electron (moving around the nucleus). The larger oxygen atom has eight protons in the nucleus, and eight electrons distributed within two shells: an inner shell with two electrons, and an outer shell with six electrons.

Outer electron shells that are completely full create a stable electronic configuration. Various elements, including hydrogen and oxygen, therefore interact with one another to form an arrangement that fills their outer shells. The outer electron shell of a hydrogen atom can hold up to two electrons, and the outer electron shell of an oxygen atom can hold up to eight electrons.

The bonding that occurs between the hydrogen and oxygen atoms in a water molecule results from two hydrogen atoms each sharing an electron with a single oxygen atom so that each hydrogen atom has two electrons in its outer shell (one of its own and one "shared"), and the oxygen atom has eight electrons in its outer shell (six of its own and two "shared"). This sharing of electrons creates a very strong bond between the atoms called a covalent bond. The resultant molecule (H_2O) is stable because all of the outer electron shells of its constituent atoms are filled. (See **Figure R1.1**.)

Figure R1.1
A water molecule is stable because the outer shells of the two hydrogen and one oxygen atoms are filled by the "sharing" of electrons.

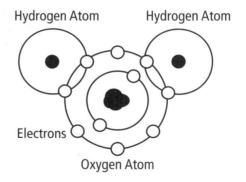

Hydrogen Atom   Hydrogen Atom

Electrons

Oxygen Atom

The angle between the hydrogen atoms in a water molecule varies from 105° in the liquid phase to 109° in the solid phase, rather than being a 180° angle as might be expected. (The increase in the angle when water freezes results in an *increase* in volume and a consequent *decrease* in density—thus, ice floats in water.) Also, the oxygen atom attracts the shared electrons more than the hydrogen atoms do. This results in an unequal distribution of electrons (and therefore negative charges) around the atoms. The oxygen nucleus attracts the shared electrons more than the hydrogen nuclei. As a result, the water molecule is electrically polarized. The hydrogen end of the molecule is partially positively charged, while the oxygen end carries a slight negative charge. (See **Figure R1.2**.) Thus, water is a polar molecule.

The electrically polarized water molecule has properties similar to a magnet—the positive end is attracted to the negative end of other molecules and vice-

versa. This electrical attraction between water molecules leads to the development
of cohesive forces called hydrogen bonds. These bonds act between two or more
water molecules, and also between water molecules and other electrically charged
particles. The degree of hydrogen bonding that occurs in water varies with its
physical state (solid, liquid, or gas), as well as with other physical properties such
as temperature.

Figure R1.2
The oxygen atom attracts
the "shared" electrons
more intensely, resulting in
a slight negative charge on
the oxygen atom and slight
positive charges on the
hydrogen atoms. Due to this
charge differential, water is
termed a polar molecule.

In ice, water molecules are bound together by multiple hydrogen bonds into a
nearly immobile crystalline structure of hexagonal shape. In liquid water, fewer
hydrogen bonds exist and the water molecules slide past one another, allowing the
water to "flow." In the vapor phase, no hydrogen bonds between water molecules
exist and water is present as solitary molecules of gas. (See **Figure R1.3**.)

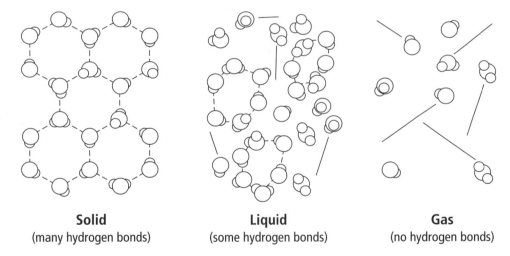

Solid
(many hydrogen bonds)

Liquid
(some hydrogen bonds)

Gas
(no hydrogen bonds)

Figure R1.3
Structure of water in solid,
liquid, and gas phases

Consider that the temperature of water is a result of the motion that its
molecules exhibit. As more energy is introduced into a collection of molecules,
the molecules move faster, hydrogen bonds are broken, and the temperature rises.
Thus, as heat energy is added to ice crystals, they vibrate more and more until
some of the hydrogen bonds are broken and the water molecules are able to move
about in the liquid state. The molecules' ability to roll and slide past one another
gives liquid water its fluid properties.

With more heat, molecules move increasingly faster until, eventually, there is
enough energy to break the remaining hydrogen bonds between molecules. The
molecules then move about randomly and have little attraction for one another;
they have entered the gaseous state and exist as water vapor.

Reading 1

To reverse the process, heat energy must leave a system containing water vapor. The speed of the molecules must be reduced to the point where hydrogen bonds form between some of the water molecules. In the atmosphere, for example, heat is released and molecular motion slows when water vapor condenses to form water droplets, as occurs in cloud formation.

As heat is removed from liquid water, increasingly more hydrogen bonds develop until the molecules are linked together in rigid hexagonal structures and ice is formed. Each of the processes that occur as water undergoes changes in state is shown in **Figure R1.4**.

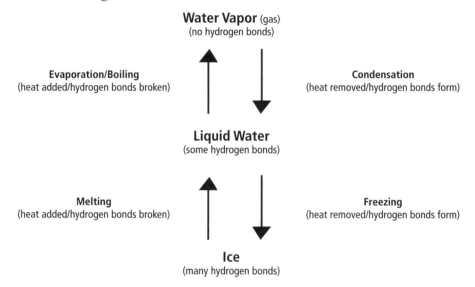

Figure R1.4
Diagram showing changes from gas (water vapor), liquid (water), and solid (ice) phases

Water is unique in that it is densest in its liquid, rather than solid, phase. The density of water is also affected by temperature. Liquid freshwater is densest at 4°C, just before hexagonal ice crystals begin to form. As the temperature of liquid water increases from 4°C, its density decreases. All other conditions being equal, warmer water is less dense than cooler water. For most other substances, the solid phase is densest and therefore sinks in the liquid phase. Obviously, this does not happen in water—ice floats. (Since water is such a common substance, many of us are more surprised that the solid phase of other substances does not float.) Ice forms at the top of a natural body of water, and stays at the top, resulting in a layer of insulation that prevents most lakes and ponds from freezing solid, even in the coldest weather. This layer of insulation also provides protection for the plant and animal life within the water beneath it. If solid water sank in liquid water, the world's lakes and ocean would tend to freeze solid.

The expansion of water upon freezing—ice occupies about 8% more space than the same amount of liquid water at 4°C (pure water's temperature at maximum density)—plays a major role in the weathering and erosion of rocks. When the water in cracks freezes, the cracks enlarge and the rock splits into fragments, which is called ice wedging.

A common misconception surrounding the expansion of water upon freezing is that air fills the spaces within the hexagonal structure of the ice crystals and is responsible for making ice less dense than liquid water. In reality, the molecules in solid water (ice) arrange themselves in a crystalline structure where there are fewer molecules per unit volume than in liquid water. (Remember that the angle between the hydrogen atoms in a water molecule increases from 105° to 109° when water freezes.) Fewer molecules per unit volume results in lower density. (See **Figure R1.5**.)

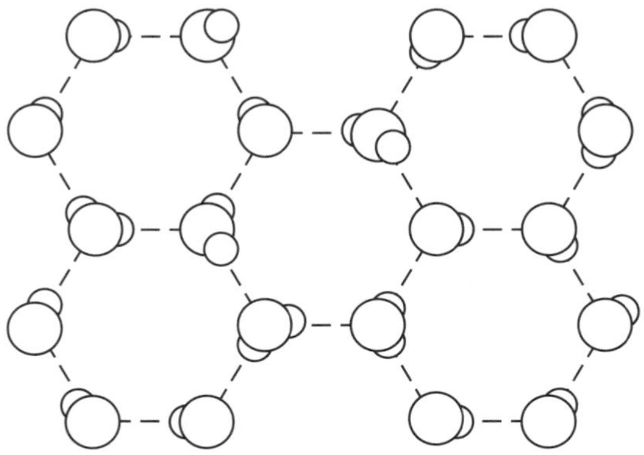

Figure R1.5
Water molecules arranged in a crystalline structure—ice

Another important property of water is its relatively high specific heat. Specific heat is defined as the amount of heat required to raise the temperature of 1 g of any substance by 1°C. Upon the addition of heat to water, hydrogen bonds must break before the temperature rises. The large number of hydrogen bonds in liquid water means that a relatively large amount of heat must be added to cause an increase in its temperature.

The high specific heat of liquid water, coupled with the large quantity of water in the ocean, allows ocean temperatures to change slowly and by small amounts throughout the year, as compared with the temperatures of the continents. The ocean is generally warmer than the continents in winter and cooler than the continents in summer. Since it absorbs and retains so much heat, the ocean moderates Earth's temperature and climate, especially in coastal regions. Also due in part to the high specific heat of water, ice sheets, icebergs, and ice floes tend to melt slowly, a factor that also contributes to making climatic change a gradual process. (They are also slow to melt because of their mass, their locations in the polar regions, and their high reflectivity.)

In addition, water is often characterized as "the universal solvent." The phenomenal ability of liquid water to dissolve is due to the polarity of the water molecule. The charged regions of the molecule interfere with the attraction between oppositely charged ions within other compounds. If the attractions between the water molecules and the ions are stronger than the bonds between the ions, the substance will dissolve in water. For example, when sodium chloride or "table salt" (NaCl) contacts water, the negative (oxygen) regions of the water molecules attract the positive (sodium) ions. At the same time, the positive (hydrogen) regions of the water molecules attract the negative (chloride) ions. The sodium and chloride ions are extracted from their crystal structure, the compound is broken down, and the salt dissolves. (See **Figure R1.6**.)

Water that contains dissolved substances has a greater weight per unit volume (density) than pure water. If 3 g of salt are dissolved in 100 ml of freshwater, the resultant mixture is 3% heavier than freshwater. The total salt content (in grams)

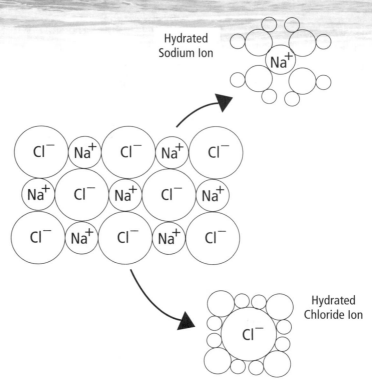

Figure R1.6
Dissolution of sodium
chloride in water

dissolved in 1 kg of seawater is referred to as *salinity*. Generally, the greater the salinity of seawater, the greater its density. Open ocean seawater has an average salinity of about 35. The convention is that salinity has no units as officially it is a ratio between the conductivity of seawater and the conductivity of a standard. Sometimes we use practical salinity units (PSU), but in reality it officially has no units.

Variation in seawater density occurs in both the open ocean and in coastal regions. In the open ocean, the variation results primarily from differences in water temperature; in estuaries and coastal zones, it results from differences in salt content. Since low-density water is more buoyant than water of higher density, low-density water "floats" at or near the surface, and water density increases gradually with depth. This density gradient in seawater bears important consequences that will be discussed in Reading 2: The Ocean.

Finally, one interesting characteristic of water is its high surface tension. Surface tension (or cohesion) is the ability of a substance to adhere to itself. Water has the highest surface tension of all liquids. High surface tension allows some insects to remain on the surface of a body of water, even though their density is greater than that of water. All known marine-dwelling insect species, in fact, live on the ocean's surface, relying on surface tension for their survival. Surface tension also plays an important role in damping out small waves on the ocean's surface. Like its other properties, water's ability to cohere is due to hydrogen bonds.

The combined effects of the various characteristics described above account for the unique nature of water. This special substance, coupled with Earth's fortunate position within the solar system, was crucial for the emergence of life and plays a vital role in allowing Earth's species to thrive.

The Ocean

Just as water is unique among molecules, its existence in our ocean makes Earth unique among the planets in the solar system. Our planet is the only body in our system that has a vast open ocean of liquid water. Of the planets close to the Sun, Mercury has at least a small amount of water in the form of ice; Venus apparently is completely dry. Very recently it was discovered that our Moon also has water present as ice. The only water found in the gas giants—Jupiter, Saturn, Uranus, and Neptune—is mixed with greater quantities of hydrogen and helium. On the smaller worlds that lie beyond Earth, including Mars, Pluto, and the moons and other satellites of all of the planets, any water that has been detected is in the form of ice. It is Earth's vast ocean, in combination with its fortunate position in the solar system, that makes our planet life-bearing.

There are two main hypotheses to explain the origin of the world ocean, which now covers over 70% of Earth's surface. The first hypothesis proposes that water vapor was slowly released from molten material beneath Earth's surface by volcanic activity. As the concentration of water vapor in the atmosphere increased, some of the vapor condensed, clouds were formed, and rain began falling on the planet's surface. The second hypothesis suggests that most of Earth's water originated in comets or carbonaceous meteorites, which were much more abundant in the first billion years of the planet's history than they are today. According to this hypothesis, as comets and meteorites entered Earth's atmosphere, they instantly vaporized, adding moisture to the air that eventually fell to Earth as rain or snow.

For whichever reason, or combination of reasons, the rain began. As it continued, the low places on Earth's surface were filled with water, and the ocean was formed. Rain, in combination with other processes such as wind, has tended to mold Earth into a smooth sphere through the processes of weathering and erosion. Fortunately for us, tectonic forces within Earth keep raising up new land masses; otherwise, our planet would eventually be covered by a vast ocean, about 2,400 m deep, unbroken by continents.

Like with any large environmental system, scientists subdivide the ocean into characteristic regions. The most fundamental scheme of subdivisions is the one followed here: coastal ocean and open ocean, with the open ocean further divided into surface, transitional, and deep layers. These oceanic subdivisions differ from one another in many ways, but treating them separately should not obscure the fact that energy, matter, and organisms move from one subdivision to the other; that is, the ocean as a whole is an integrated environmental system, even though scientists subdivide it for convenience of study. (See **Figure R2.1.**)

Reading 2

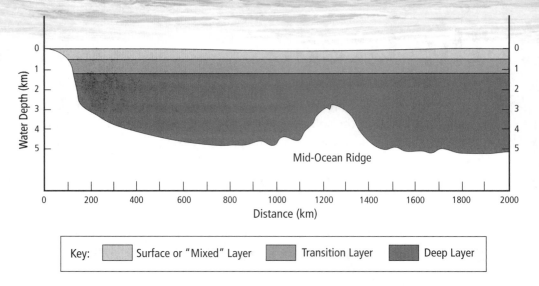

Figure R2.1
Simplified subdivision
of the ocean

Key: ▢ Surface or "Mixed" Layer ▢ Transition Layer ▢ Deep Layer

The coastal ocean refers to the 10% of ocean over the continental shelves (the undersea extensions of the continents, which descend gradually and are not part of the deep ocean floor). The coastal ocean provides almost all of the world's seafood harvest, it is the source of all of the minerals and petroleum currently recovered from the ocean, and the coastline itself is home to a large and growing percentage of the human population. According to the National Oceanic and Atmospheric Administration, more than half of the population of the United States lives within 50 miles of the ocean or Great Lakes coastlines. The concentration of natural resources, human population, and economic activity make it no surprise that almost all conflicts and controversies regarding the use of the ocean originate from coastal ocean activities.

The single most characteristic feature of the coastal ocean environment is rapid change. Coastal winds are strong and variable, and they drive much of the coastal water circulation. Coastal winds also create waves that break on shore and move sediments, an action that can contribute to beach erosion and expansion. Variations in air temperature are generally reduced by the capacity of ocean water to store and release heat; but local and regional winds can overwhelm the moderating effect at relatively low speeds. Humidity, especially in summer, is generally high, and helps fuel potentially violent thunderstorms and hurricanes that can quickly alter human activities and environmental conditions. Freshwater and sediment runoff from major rivers—especially after prolonged periods of rainfall—can also dramatically affect the coastal environment. The complexity of the coastal environment may lead to oversimplification of ideas or the development of preconceptions among students.

An important feature of the coastal ocean is the estuary. An estuary is a semienclosed body of coastal water that exhibits measurably reduced salinity due to the introduction of freshwater from rivers, streams, and other sources of continental runoff. Most estuaries exhibit two-way movement of water. Less dense fresh (river) water flows seaward along the surface, while denser, salty ocean water flows landward underneath it. Varied amounts of mixing of freshwater and salt water occur, depending on factors such as wind speed, tidal range, the depth and contour of the estuary bottom, and the relative inflow of river water versus seawater. When mixing is minimal, a salt-wedge type of estuary develops, with a wedge of seawater undercutting an overlying wedge of freshwater. (See **Figure R2.2**.)

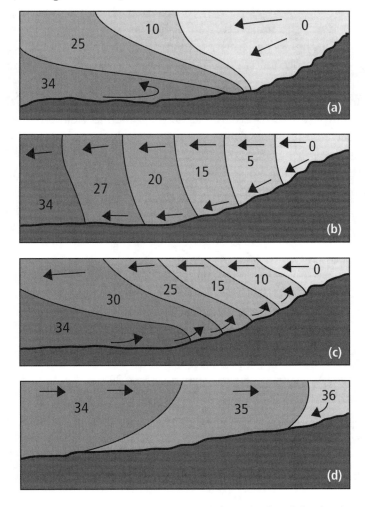

Figure R2.2
Schematic of the vertical cross section along the axes of four types of estuaries (from a physical oceanography perspective): (a) salt wedge, (b) well-mixed, (c) partially mixed, and (d) reverse. Salinity values in each type show the amount of mixing between freshwater (0 salinity) and seawater (35 salinity). Solid lines represent lines of equal salinity. The arrows indicate flow direction, and the thick black line is the bottom of the estuary.

Estuaries and other coastal regions account for much of the biological productivity of the ocean. These areas are nutrient-rich from the accumulation of materials brought from the land via runoff. They also provide relatively sheltered habitats that are the spawning places and nurseries for many forms of marine life.

The remaining 90% of the ocean is open ocean—that is, it occurs seaward of the world's continental shelves. The open ocean is characterized by the surface layer, the transitional layer, and the deep water zone. In most areas, turbulence due to wind, waves, and surface currents produces fairly constant physical properties (temperature and salinity) in the surface—or "mixed"—layer. In general, this layer is a few hundred meters thick (about 300 m in equatorial regions); its depth varies according to latitude and frequency of storms. The mixed layer acts as a thermal cap.

Some of the most predominant winds that stir the currents are the trade winds and prevailing westerlies. These global winds blow in opposing directions and, in combination with Earth's rotation, cause a powerful system of rotating currents—

called gyres—to develop. The water in the Gulf Stream region of the North Atlantic Gyre, for example, moves at speeds of up to 112 km/day (3 knots). The correlation between global wind patterns and surface ocean currents can be seen in **Figure R2.3**.

Below the surface zone, the temperature drops from whatever it is at the surface to the average temperature of deep water, which is a few degrees Celsius. This region of change is called the transition zone and is where we see the presence of a thermocline (rapid change in temperature), halocline (change in salinity), and pycnocline (change in density). The transition zone has a strong thermocline in equatorial regions where the difference between the surface and deep water temperatures is greatest. The thermocline weakens with increasing latitude, and weakens seasonally in winter months in the mid-latitudes. On average, the transition zone extends to a depth of about 1,000 m.

Beneath the transition zone, the physical properties of water are quite stable and uniform. This is the third layer, the deep water zone.

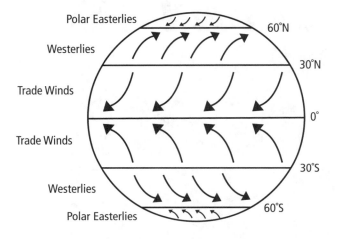

Global Wind Circulation Patterns

Figure R2.3
The relationship between global wind patterns and ocean surface currents

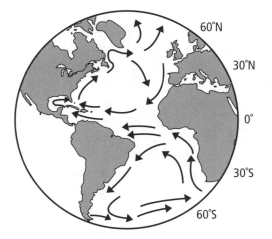

Surface Ocean Currents in the Atlantic Ocean

The slowly circulating open ocean deep layer covers more than 75% of the area of the ocean basins, and more than half the total surface area of Earth. The deep ocean layer is the area of Earth about which our understanding is uncertain. We have determined, however, that the waters of the deep sea circulate in a pattern of convection currents that results from differences in the densities of masses of seawater. As discussed in Reading 1: Water: The Sum of Its Parts, the density of water increases with an increase in salt content (salinity) or a decrease in temperature; and denser water tends to sink relative to less dense water. Therefore, seawater with high salinity sinks relative to seawater with lower salinity, and cold water sinks relative to warmer water.

Much of the density variation in the deep ocean layer has its origin in the polar regions of Earth where seawater in the surface layer freezes. When water freezes to form ice sheets, which in turn sometimes break apart to form icebergs, the salt it contained remains dissolved and is concentrated in the remaining unfrozen ocean water, increasing its salinity. (This is because the salt ions cannot easily fit into the crystalline structure of the ice.) In addition, the concentration of salts in the unfrozen water lowers its freezing point, allowing it to remain liquid below its normal freezing point of 0°C. This cold, salty water is denser than the surrounding surface water and, therefore, it tends to sink toward the seafloor and move away from the poles. Elsewhere, deep water is displaced upward toward the surface layer.

An example of a dense, deep ocean water mass is the Antarctic Bottom Water (AABW). The densest water mass in the ocean, the AABW is created when ice forms in the ocean around Antarctica, leaving behind extremely salty water with a temperature of less than 0°C. This water sinks to the seafloor as a coherent and identifiable mass. It creeps along the bottom of the ocean basin at speeds as slow as 30 m/day. Contrast this speed with the rapid 112,000 m/day movement of the Gulf Stream surface current! The AABW has been traced as far north as Bermuda and France before converging and mixing with a much larger water mass from the North Atlantic. (See **Figure R2.4**.)

It is important to note that the convection currents that occur in the deep ocean layer are perpetuated by the freezing of surface water into ice near the poles. This leads to an important aspect of convection that should be pointed out to students: convection currents are not necessarily caused by the introduction of heat. Rather, they are caused by density variations within a fluid. Variations in density can result from changes in temperature (due to uneven heating or cooling), salinity, or both. The deep ocean layer's system of density-driven flow (convection) is referred to as "thermohaline" circulation, since it is caused by differences in temperature (thermo) and salt content (haline) of oceanic water masses.

The concept of a three-layered ocean is important to the study of the ocean as a whole. It is also an essential feature of all mathematical models of Earth's climate, since the surface layer exchanges water and energy with the atmosphere, while the deeper layers do not. It is important to recognize that the layers are not totally distinct and that individual water molecules circulate throughout both. It may take hundreds of years, however, before cold, dense ocean water that sinks to the ocean floor near the poles returns to the surface layer again.

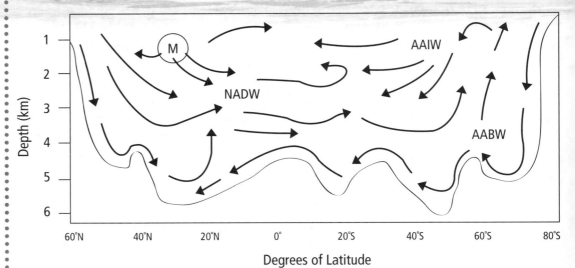

Figure R2.4
Deep ocean currents in the Atlantic Ocean

Key

NADW = North Atlantic Deep Water

AAIW = Antarctic Intermediate Water

AABW = Antarctic Bottom Water

M = Inflow of Water From the Mediterranean

The Tides: A Balance of Forces

Despite the fact that tides are a phenomenon familiar to most individuals, the driving forces mystify and confuse many people. Preconceptions about what causes tides abound, and most people have only a vague awareness that the Moon is somehow involved. A global view of the tides can help explain the complex actions and interactions of the forces involved in tide formation.

You can think of tides as two crests of water (or tidal bulges), and two troughs (or tidal troughs) midway between the crests, in a global ocean. Each crest and trough runs the length of Earth from the North Pole to the South Pole. One bulge is always oriented toward the Moon, and the other is always oriented away from the Moon. The bulges (and troughs) shift position following the Moon's movement around the center of mass of the Earth–Moon system. As the speed of this movement is much slower than Earth's rotation speed, the bulges appear almost stationary, and this view is therefore called the static model of tides. **Figure R3.1** represents a simplified version of the static model.

To observers on Earth, the tidal bulges are experienced as high tides, and the tidal troughs as low tides. The tides change as Earth rotates. Since it takes 24 hours for Earth to make one complete rotation (360°) on its axis, someone may suggest that consecutive high or low tides would occur exactly 12 hours apart (the time it would take for Earth to complete one-half a rotation (180°) from one tidal bulge or trough to the other). In reality, the period between consecutive high or low tides is approximately 12 hours, 25 minutes. The reason for the difference is because the Moon is moving relative to

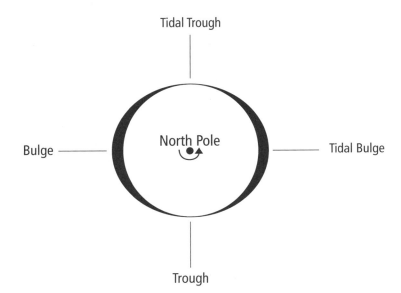

Figure R3.1 Top view diagram showing tidal effects on Earth under the static model

Reading 3

Topic: tides
Go to: *www.scilinks.org*
Code: PESO 011

Earth—it is revolving around Earth. As a result of this movement and its effect on the position of the ocean's bulges and troughs, it takes longer than 12 hours for a specific location on Earth to rotate from directly beneath a tidal bulge or trough to the next bulge or trough (i.e., the bulges and troughs have moved).

An analogy can help provide a clearer picture of this situation. Imagine that two people—a runner and a walker—are standing still at the same point on an oval, quarter-mile track. The runner can run around the track in one minute; the walker goes much more slowly. They both start around the track at the same instant and in the same direction—one running, the other walking. In one minute, the runner will be back at the starting point; but he will not have caught up with the walker, who has also moved around the track from the starting point. More time will be required for the runner to catch back up with (or "lap") the walker.

Similarly, on Earth, a location that experiences a high tide at 2:00 a.m. will rotate halfway around Earth in 12 hours. During that time, the bulge will have rotated to a much smaller degree in the same direction. Some extra time (25 minutes) will therefore be required for the location to move beneath the other tidal bulge and experience its second high tide. High tide will occur at 2:25 p.m. rather than 2:00 p.m. At 2:00 a.m. the next day, the location will have made one complete rotation, but the tidal bulge on that side of Earth will have continued to rotate away from its original position. The location will therefore not experience its next high tide until 2:50 a.m., 12 hours and 25 minutes after its previous high tide or 24 hours and 50 minutes after its initial high tide. (See **Figure R3.2**.)

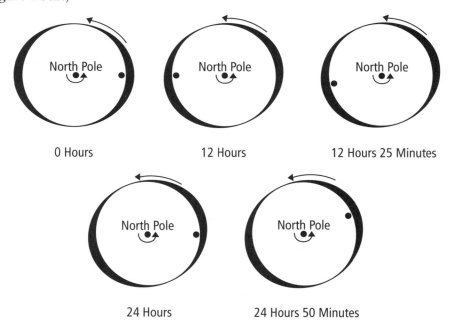

0 Hours 12 Hours 12 Hours 25 Minutes

24 Hours 24 Hours 50 Minutes

Figure R3.2
Rotation of Earth and of the tidal wave
Note: The dot along the edge of Earth in the figure represents the position of the observer.

Two concepts from physics are helpful in explaining tides. Newton's law of universal gravitation states that any two objects are attracted to each other by a force—gravity—that is directly proportional to the product of their masses and inversely proportional to the square of the distance between them. The greater the mass of an object, the greater its gravitational attraction; the greater the distance from an object, the smaller the effect of its gravity. Newton's law of inertia states that a body at rest tends to stay at rest, and a body in motion tends to stay in motion, in a straight line, unless acted upon by an outside force. These two phenomena—gravity and inertia—work together to create the tides.

According to Newton's law of inertia, the Moon, as it moves through space, has a tendency to continue its motion in a straight line and bypass Earth. However, according to Newton's law of universal gravitation, Earth and the Moon are attracted to one another due to gravity. Since Earth has much greater mass than the Moon, the effect of Earth's gravity is stronger. It acts in conjunction with the Moon's inertia to keep the Moon in orbit around Earth. Without the effect of Earth's gravity, the Moon would fly off in a straight-line tangent to its orbit around Earth. (See **Figure R3.3**.)

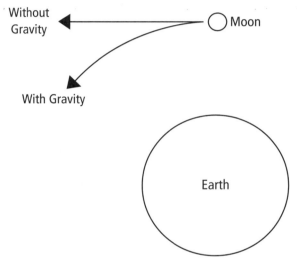

Figure R3.3
Moon's trajectory with and without gravity

The Moon also exerts a gravitational attraction on Earth. Land masses and the ocean are both affected by the Moon's gravity, but because water is more easily deformed than land, land masses are affected to a lesser degree and ocean water moves over the area of Earth facing the Moon. In this way, the Moon's gravity causes the tidal bulge that is oriented toward the Moon. (See **Figure R3.4**.)

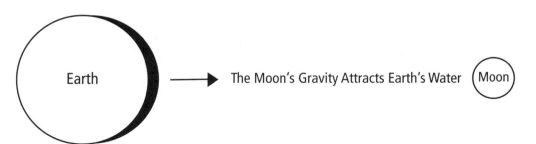

Figure R3.4
The Moon's gravity attracts Earth's water.

People sometimes think that the center of the Moon's orbit around Earth is in the center of Earth where Earth's center of mass and axis of rotation are located. In reality, Earth and the Moon together form a two-body system that rotates on an axis located at the center of mass of the system. (Students might have learned this in Activity 16.)

The center of mass of the Earth–Moon system is distinctly different from Earth's center of mass. It is located within Earth, but not at the center of Earth. This is because Earth is so much more massive than the Moon. The analogy of a seesaw can help make this point clearer. If an adult and a small child are on a seesaw, the adult must sit much closer to the fulcrum in order for the seesaw to be balanced. Similarly, if Earth and the Moon were placed on a seesaw, the fulcrum would have to be located within Earth for the two to balance. (See **Figure R3.5**.)

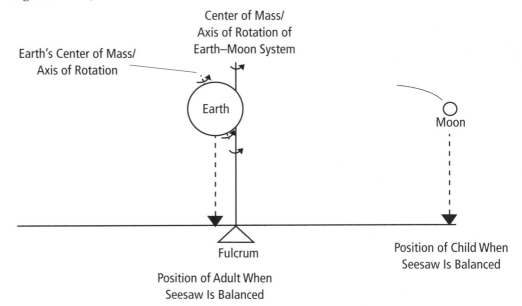

Figure R3.5
Center of mass of Earth and of the Earth–Moon system and its analogy with a seesaw

As the Earth–Moon system rotates about its axis, Earth, like the Moon, has an inertial tendency to continue in a straight line through space. Everything on the planet, including its ocean, is also subject to the effect of inertia. While Earth's gravitational attraction for the ocean keeps it from flying off the planet, it does move away from Earth somewhat because of its fluid nature.

This is what creates the tidal bulge on the side of Earth opposite the Moon. The bulge is shown in an exaggerated way in **Figure R3.6**.

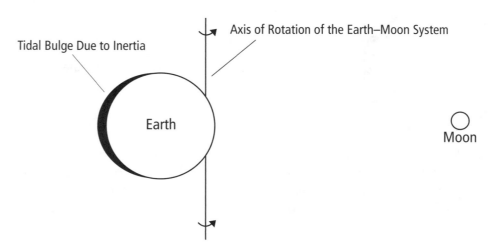

Figure R3.6
Tidal bulge caused by inertia

The inertia of the Earth–Moon system contributes to the tidal bulge on the Moon side of Earth as well, but to a much smaller degree. Since the Moon side of Earth is very close to the center of mass of the Earth–Moon system, the force of inertia there is much less. Activity 15 explores this difference.

In summary, the two tidal bulges on Earth result, for the most part, from two different factors. The bulge on the side of Earth facing the Moon is due primarily to the effect of the gravitational attraction of the Moon on Earth's water. The bulge on the opposite side of Earth is due primarily to the inertial tendency of water to travel in a straight line away from Earth as the Earth–Moon system rotates. (See **Figure R3.7.**)

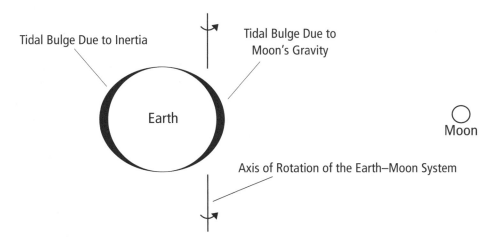

Figure R3.7
Earth–Moon system showing the two tidal bulges

The tidal bulges associated with the rotation of the Earth–Moon system should not be confused with the equatorial bulge that results from the rotation of Earth on its own axis. The equatorial bulge, although caused by inertia, is not a tide. It is a constant effect that occurs uniformly around the planet's equatorial regions regardless of the position of the Moon. (See **Figure R3.8.**)

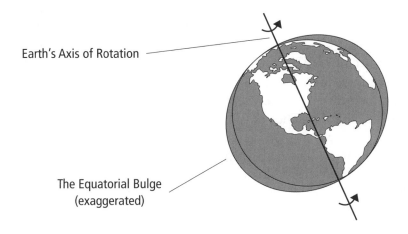

Figure R3.8
Earth showing the equatorial bulge

The ocean is also affected by the gravitational attraction of the Sun. While the effect is less than half that of the Moon's, the Sun's effects become most evident roughly four times a month—approximately at new moon, first quarter, full moon, and third quarter. At new moon, the Sun's gravitational attraction works in conjunction with the Moon's to cause an exceptionally high tide. (Students can explore this in Activity 15.)

The high tide associated with a new or full moon is called a spring tide because the ocean "springs" away from Earth even more than usual. At new and full moon, the Sun's gravitational attraction adds to the bulge on the side of Earth opposite the Moon, creating a second spring tide. At both first and third quarter, the Sun's gravitational attraction pulls water away from the lunar tidal bulges and diminishes them. This creates the moderately high and low tides referred to as neap tides. Spring and neap tides each occur twice every 29.5 days. (See **Figure R3.9**.)

Coastal locations experience tides as rhythmic fluctuations in sea level. At high tide, the sea level rises; at low tide, the sea level falls. Not all locations experience the same pattern of tidal activity, however. Most locations in North America, for example, experience either semidiurnal tides—two high tides and two low tides every 24 hours and 50 minutes—or mixed tides, where there is a *high* high tide, a *low* high tide, a *high* low tide, and a *low* low tide every 24 hours and 50 minutes. Locations on the Gulf of Mexico have a diurnal or daily tide—one high tide and one low tide every 24 hours and 50 minutes.

Variations in tidal patterns result from the fact that the depth of the ocean is not uniform. Water in coastal ocean basins oscillates at frequencies either operating in conjunction with or opposing those of the basic tide-producing forces. Just as the water in a bathtub sloshes back and forth at a frequency characteristic of its shape and bottom contour, the ocean water in coastal basins "sloshes" back and forth at a frequency unique to each basin. The coastal basin that affects most of North America's eastern coastline has an oscillation period of about 24 hours and 50 minutes, and therefore tends to reinforce tidal action. The reinforcing action is best seen in Canada's Bay of Fundy, where oscillation and tidal periods reinforce each other to produce differences in sea level of up to 18 m between high and low tides. Off the coast of California, the coastal basin oscillation reinforces every other tide, thereby creating the *high* high and *low* low tides. In the Gulf of Mexico, the period of oscillation is every 24 hours, thereby creating a single high tide and a single low tide each day.

A common misconception about tides is that centrifugal force is partly responsible for them. The concept of centrifugal force is a myth. It is described as a force that pulls an object out away from the center of its circular orbit. The tidal and equatorial bulges are attributed to this nonexistent force through the assertion that water is being pulled away from Earth as Earth rotates. While the assertion that water tends to move away from Earth is true, the conclusion that centrifugal force causes the tendency is incorrect. If centrifugal force were at work, the water would tend to move along a radial path away from Earth; but the real tendency of water is to move along a path tangent to Earth. This tendency results from inertia, as previously described. Centrifugal force is not

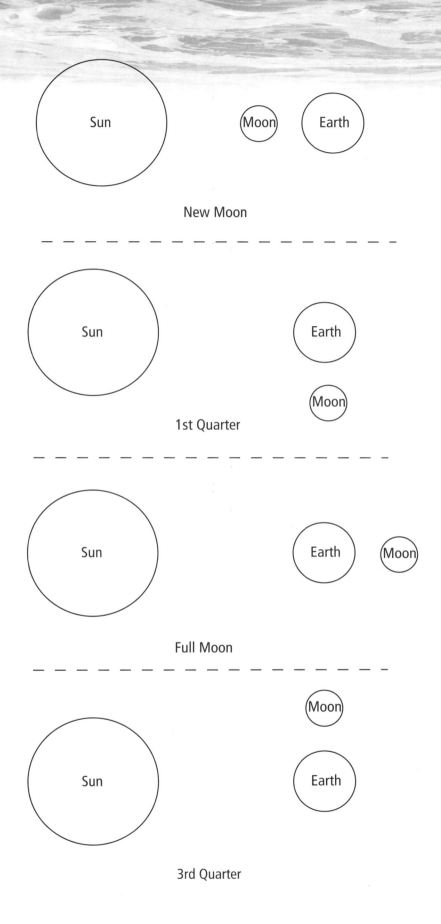

New Moon

1st Quarter

Full Moon

3rd Quarter

Figure R3.9
Schematic position showing the relative position of the Sun, Earth, and Moon during the different phases of the Moon

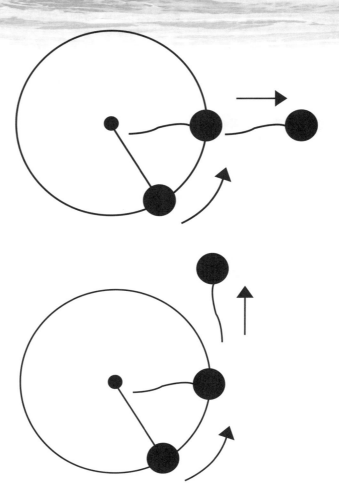

Figure R3.10a
Incorrect trajectory of the ball after the string breaks under the "mythical" centrifugal force

Figure R3.10b
Correct trajectory of the ball after the string breaks under realistic inertial conditions

a real force, but a perceived effect. Some physics writings dispel the myth of centrifugal force and point out how it is misleading. (See **Figure R3.10.**)

In teaching students about tides, it is important for you to address misconceptions such as centrifugal force. You can show the difference between centrifugal force and inertia with a simple demonstration. Give a student a ball attached to a string and ask the student to swing it about his or her head like a sling and to aim it at a target directly in front of him or her. The student will soon discover that centrifugal force is indeed a preconception. If the student releases the string when the ball is directly in front of him or her (as centrifugal force suggests is necessary to hit the target), the ball will miss the target completely. With practice, the student will learn when to release the string so the ball's tangential path will lead to the target.

In summary, while the tides are sometimes attributed to the effect of centrifugal force, this force is a preconception. The tides are actually generated by complex interactions of the forces of gravity and inertia inherent in the rotating Earth–Moon system. They can be viewed as a global wave under which Earth rotates, with crests experienced as high tides and troughs as low tides. The tides are reinforced or diminished by factors such as the natural oscillation inherent to coastal basins and the gravitational effect of the Sun.

Waves

Ask anyone to name a characteristic feature of the ocean and, most likely, their answer will relate to waves. Waves figure significantly in our view of the ocean because breaking waves are such a prominent feature along the beach. A careful review of wave characteristics, ocean wave formation, and types of ocean waves should address some of the natural curiosity, as well as some common preconceptions, that many students have about waves.

A wave is a disturbance that transmits energy from one place to another. This definition applies to light and sound waves as well as ocean waves. Terms commonly used to describe waves of all types include the following: crest—the highest point of a wave; trough—the lowest point of a wave; wave height—the vertical distance from crest to trough; wavelength—the horizontal distance between two consecutive crests (or two other similar points); frequency—the number of crests that pass a fixed point per unit of time; and finally, speed—the distance a wave travels per unit of time. In addition to wave height, wave period is commonly reported by NOAA buoys. Wave period information is sought by surfers to gauge the magnitude of storm swell. Higher period waves are the most desired by extreme wave surfers. The wave period is the time between the wave crests. All three factors—wind speed, duration, and fetch—influence the height and wavelength of waves. (See **Figure R4.1**.)

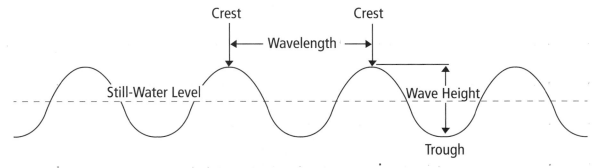

Figure R4.1 Parts of a transverse wave

There are a number of different types of ocean waves, the most familiar generally being those that break onto the beaches. Beach breakers are generated by the force of the wind. There are three characteristics of wind that influence the formation of a wind-generated wave: its duration (the length of time it has been blowing), its fetch (the distance over which the wind is blowing), and its average velocity (or speed). Fetch is related to the wavelength of the waves, while increases in the velocity and duration of the wind cause increases in the wavelength and height of the waves.

Reading 4

A common preconception among students is that the water that rushes ashore as a wave breaks is water that has traveled with the wave from many miles off shore. This, however, is not the case. As a surface wave moves through water, it is only the wave energy or waveform that advances; the water particles (molecules) through which the wave passes move in a circular pattern and exhibit little or no overall forward motion. As the crest of a wave encounters a water particle, the particle moves to the peak of its path. As the crest passes by, the particle continues to the bottom of its circular path. (See **Figure R4.2**.)

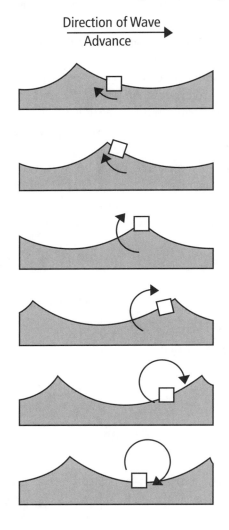

Figure R4.2
Water particle movement in a surface wave

Wind waves form because, as wind blows along the surface of a body of water, air pressure fluctuations and friction drag water into a series of crests and troughs. In the area directly affected by the wind, the waves, called sea waves, are irregular and travel in different directions. As sea waves move out from the area where they originate, longer waves with higher velocities eventually overtake the shorter, slower waves. The long and short waves combine to form a regular pattern and become swell waves. The energy in a swell wave affects the water through which it passes to a depth equivalent of one-half the wavelength of the wave. Below this depth, water and the ocean bottom are unaffected by the wave's energy.

Swell waves can cross the ocean if they find water of sufficient depth. Once a swell wave moves into water roughly one-half of its wavelength or less in depth, however, the waveform encounters and begins to "feel" the bottom (seafloor). When a wave "feels" the bottom, its height increases and its velocity decreases. Due to the friction of water particles against the seafloor, the top of the wave moves ahead faster than the bottom of the wave. The wave eventually forms a curl of water that falls over forward, creating a breaking wave. (See **Figure R4.3.**)

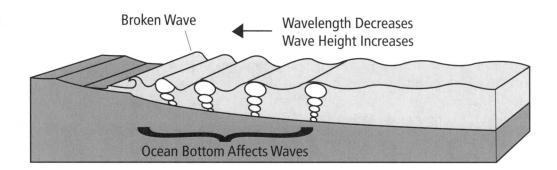

Figure R4.3
Effect of the bottom on waves

If waves enter shallow water while moving at an angle to the beach or encounter irregular bottom features, the waves change direction—they are refracted. The part of a wave that first encounters the bottom will slow compared to the rest of the wave, causing the entire wave to turn (refract) toward the shallow water. In this way, waves tend to parallel shoreline contours. They converge on areas of the shore that protrude, and diverge from areas of the shore that curve landward. (See **Figure R4.4.**)

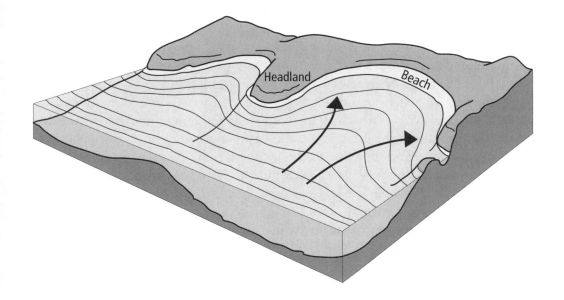

Figure R4.4
Refracted waves parallel the shape of the coastline.

Water waves represent energy in motion. When a wave breaks, its energy is transferred to obstacles it encounters along the shoreline. The beach, which acts as a barrier to breaking waves, receives the brunt of their energy. Waves continually provide energy for the transport and redistribution of large amounts of sand through the shallow coastal waters. The energy released from waves often leads to the development of local water currents. Waves that approach the beach at an angle, for example, can transport water and sand parallel to the beach (a longshore current), or perpendicular to the beach (a rip current).

Ordinary wind-generated waves cause continual, gradual changes in beach profiles over time through natural processes of sand deposition and erosion. Large, extremely powerful waves such as storm surges and tsunamis, however, can spread immediate devastation over large areas of the coastal environment. The main cause of storm surges is high winds pushing on the surface of the ocean, which results in water piling up against the shoreline. Low pressure at the center of a storm also has a small secondary effect, resulting in high water level in the center of the storm.

Tsunamis, on the other hand, are great sea waves produced by the sudden displacement of a large volume of ocean water. Some causes of these sudden displacements are undersea earthquakes, landslides, and violent volcanic eruptions. As tsunamis travel through deep ocean water, their wave heights rarely exceed 30 cm, yet their wavelengths range from 120 to 720 km and their velocities in areas of deep ocean exceed 600 km/h! Due to their long wavelengths, tsunamis always "feel" the bottom of the ocean, but they begin to grow in height and shorten in wavelength dramatically as they near shallow coastal water. As with wind-generated waves, when a tsunami comes in contact with the bottom of the ocean, the bottom of the wave slows. The top of the wave piles up a wall of water that brings devastating effects when it breaks on shore. Regions of the world at risk from tsunamis have developed a warning system to try to alert residents in the event of approaching danger. (The term *tsunami* is borrowed from the Japanese and translates to "harbor wave." The term *tidal wave* is often used interchangeably with tsunami; however, a tsunami is not shaped by the same forces as tides. Therefore, the use of the term *tidal wave* is inappropriate.)

Tsunamis can be devastating, as the world saw in 2004 in Indonesia and in 2011 in Japan. With the latter, an extensive tsunami warning system was in place. Unfortunately, the earthquake and ensuing ocean floor displacement were immense and occurred too close to the Japanese coast for tsunami warnings to save thousands of human lives. (Reading 3: The Tides: A Balance of Forces discusses tsunamis within the broader context of earthquakes.)

Tides are also water waves, even though ordinarily they are not experienced in the same way as waves caused by wind. An extensive review of tide formation may be found in Reading 3: The Tides: A Balance of Forces.

The Ocean: A Global View

The world ocean plays an integral role in the lives of all of Earth's inhabitants through its effect on global climate. As explained in Reading 1: Water: The Sum of Its Parts, the ocean's significance in climatic regulation is largely attributable to various characteristics of water. The high specific heat of water means that a relatively large amount of heat energy must be absorbed or released by liquid water before a change in its temperature can occur. Due to the large volume of water in the ocean, ocean temperatures change slowly and vary less than the temperatures of landmasses.

The ocean helps moderate temperatures worldwide by affecting the conditions of air masses that move across the water surface. Air masses, when they move over land, moderate the temperature of landmasses through precipitation and heat exchange. The moderating effect is most obvious in coastal regions, but regions far from any coastline also experience this influence. In fact, Earth's ocean causes the planet's overall temperature to be relatively consistent over time, especially when compared to that of other planets. On other planets, and even on Earth's own Moon, surface temperatures fluctuate much more widely during the course of a year, and in some cases, between day and night. Earth's relatively consistent temperature, maintained largely by its vast ocean, distinguishes it from its neighbors in the solar system and helps make it a habitable planet.

The ocean also affects global climatic conditions as a key component of the hydrologic cycle—the interactive cycling of water between the ocean, continents, and the atmosphere. In the hydrologic cycle, heat energy from the Sun causes liquid water to evaporate and be introduced into the atmosphere as water vapor. Evaporation takes place from the surface of all bodies of water, as well as from soil and living organisms. Water vapor rises into the atmosphere, cools, and condenses to form the tiny water droplets that combine to form clouds. When the amount of water in the atmosphere reaches a critical level, precipitation occurs and some of the evaporated water is returned to Earth. Since the ocean covers over 70% of Earth's surface, most precipitation falls there and mixes with seawater. Precipitation that falls onto land either evaporates directly back into the atmosphere, is absorbed by the soil, or flows into streams. Small streams combine to form larger streams and rivers, which eventually flow into a lake or the ocean. All along the way, water is reintroduced into the atmosphere through evaporation, and the cycle continues. (See **Figure R5.1.**)

Figure R5.1
Global Water Cycle

Water is the only substance that occurs naturally on Earth in all three states: solid (ice and snow), liquid (ocean, river, and lake water), and gas (clouds, steam, and vapor). The natural circulation or pathway that Earth's water follows as it changes between liquid, solid, and gas states is called the global water cycle.

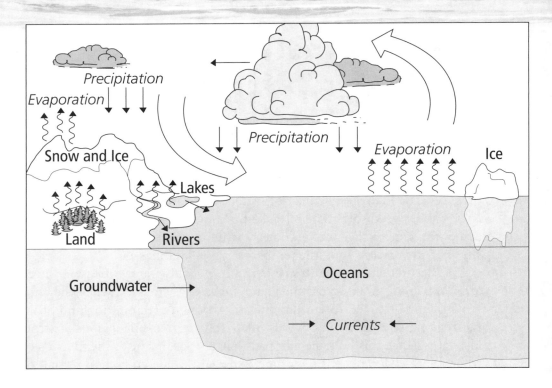

Topic: water cycle
Go to: *www.scilinks.org*
Code: PESO 001

The salinity of the ocean, important for the survival of its inhabitants, is maintained through the hydrologic cycle. When rainwater falls on land, it erodes and dissolves salts and minerals from rocks and soil. These salts are carried by runoff through streams and rivers, and many eventually reach the ocean. It is estimated that some four billion tons of dissolved salts are carried into the ocean each year. At the same time, water that evaporates from the ocean leaves its salts behind, increasing the salt concentration of the remaining seawater. The ongoing introduction of salts into the ocean through these processes is offset by other events. Freshwater continually enters the ocean as precipitation, and salts are deposited on the ocean floor at about the same rate as they are introduced into the ocean by rivers and streams. The balance between these opposing processes has caused the salinity of ocean water to remain essentially the same through recent geologic history.

Within the last few decades, scientists have developed a fairly clear understanding of the role the ocean and the hydrologic cycle play in weather formation. One interesting result has been the explanation of the effects of El Niño. El Niño is a complex interaction between the atmosphere and ocean that results in a reversal of prevailing trade winds and a rise in the temperature of the surface waters of the equatorial Pacific as the ocean currents associated with the upwelling of cold, nutrient-rich waters from the deep ocean are weakened. The warm surface water temperatures associated with El Niño lead to alterations in the movement of air masses and the development of irregular ocean currents. Changes in the normal formation of weather conditions result, and the entire world is eventually affected. In the mid-1980s, it was found that El Niño was responsible for weather extremes, from severe drought conditions in Australia and Africa to heavy rains and flooding in parts of South America. More recently, scientists have found that El Niño impacts the eastern United States (generally wetter winters), and the Caribbean basin (generally warmer and drier winters) as well. Also, El Niño periods have been found to result in much less active hurricane seasons in the Caribbean.

The effects of El Niño provide evidence of the integral role the ocean plays in the lives of human beings. As marine explorer Jacques-Yves Cousteau stated, "The very survival of the human species depends upon the maintenance of an ocean, clean and alive, spreading all around the world. The ocean is our planet's life belt." The importance of the ocean is often overlooked, however, even as people around the world alter their activities in response to daily weather conditions regulated in large part by the ocean.

Similarly, the important impact that humans have on the ocean is generally not fully acknowledged. The pollution that results from certain human activities has the potential to permanently alter the delicate ecological balance in the ocean and thereby undermine its capacity to sustain life. Human impacts on the ocean include global warming, sea level rise, ocean acidification, overfishing, eutrophication, habitat destruction, and the introduction of invasive species.

One familiar ocean contaminant is oil. Oil seeps, in which crude oil exudes from the ocean bottom, are common occurrences in some areas. Natural degradation processes occur after an oil seep that, given time, generally bring about recovery of the affected area. Ocean pollution, on the other hand, arises from human activity that introduces an "unnatural" amount of oil, thereby overwhelming the capacity of the environment's degradation and recovery mechanisms. Oil enters ocean water primarily during transportation activities. While in transport, oil may be released by "normal" leakage, accidents, or improper equipment cleaning methods. Oil may also spill into the marine environment during oil exploration or production, as was the case with the 2010 BP Deep Water Horizon (generally referred to as the Gulf Oil Spill) incident. Almost 5 million barrels (estimated at 53,000 barrels/day) of oil gushed into the northern Gulf of Mexico, making it the worst oil spill of all time.

Immediate damage to marine life results from an oil spill, which generally occurs at the interface between the ocean and the atmosphere, where the oil forms a layer or "slick." As the *Exxon Valdez* accident of 1989 and the 2010 BP spill have clearly demonstrated, seabirds, sea otters, and other aquatic animals that must surface for air are extremely vulnerable to the effects of spilled oil. The oil adheres to their feathers, skins, and coats, and severely limits their mobility. They often become easy prey for predators or starve because they can no longer hunt successfully. Those that survive and are rescued must be individually cleaned and treated.

Less obvious but possibly long-term problems can result from toxic hydrocarbon components of oil. As oil that remains in the marine environment breaks down into its various components, some of the more volatile compounds evaporate into the atmosphere and others decay relatively quickly through natural processes. Other compounds, however, dissolve into the water or settle to the ocean bottom where they can enter the food chain if ingested by organisms. This has been the case with the BP Deep Water Horizon disaster. Through the food chain, these toxins are concentrated and their destructive effects are multiplied over time. Humans, at the top of the food chain, may be affected through this route.

Several techniques are used to contain and recover spilled oil. (Students model these methods in Activity 17.) One technique involves specially equipped boats that skim the oil from the ocean's surface; workers can place floating booms around an oil slick in an attempt to contain it; they can spray chemical dispersants over an oil slick to break down the oil and cause it to degrade faster; and, at times, they even ignite the oil. A new technique developed involves oil-eating bacteria. In many cases, however, efforts to clean up an oil spill meet with only limited success, and some cleanup techniques can pose greater threats to the environment than the oil spill itself. Ignition of an oil spill, for example, causes air pollution; and chemical dispersants cause clumps of toxic material to sink to the ocean bottom where they are likely to be consumed by oysters, clams, and crabs.

Spilled oil that is not contained before it reaches shorelines can spoil beaches and thereby threaten the habitat of shorebirds and other beach-dwelling creatures. Where possible, workers absorb pools of oil with hay; they scrub rocks and other objects by hand.

The *Exxon Valdez* incident of 1989 led to the Oil Pollution Act, an amendment to the 1973 Clean Water Act that requires some oil storage facilities to prepare Facility Response Plans. In 1994, the EPA finalized the revisions that direct facility owners or operators to prepare and submit plans for responding to a worst-case discharge of oil. Following the 2010 BP incident, new legislation is being enacted concerning prevention, preparedness, response, liability and compensation, and restoration.

Ocean dumping (the disposal of waste products at sea) is another major form of ocean pollution. The shallow waters off the coasts of many countries are often used as dumping grounds for various wastes. The choice to dump garbage at sea, as opposed to a landfill, is usually based on economic reasons, as land is generally deemed more valuable than the ocean.

Among the many pollutants disposed of through ocean dumping are a wide variety of plastics. In 2007, 260 billion kg (287 million tons) of plastic were produced worldwide, with annual growth of about 3% in U.S. production since 1980. Packaging accounts for 37% of worldwide production. In North America and western Europe, per capita consumption in 2007 was 100 kg. During a subsequent National Coastal Cleanup, 62% of the trash collected was plastic. In a 1998 survey, plastic made up 89% of the trash observed floating in the North Pacific Ocean. Estimates are that an average of 13,000 pieces of plastic float per square kilometer of ocean. In one day in 2009, 380,000 volunteers collected 2,053,768 plastic bags, plastic water bottles, and six-pack holders on ocean beaches. Over the course of 25 years of beach cleanups, volunteers have collected 7,825,319 plastic bags.

Obviously, one of the problems inherent in dumping plastics at sea is that most of it floats. Another problem is its durability. Plastics are relatively immune to the natural process of decomposition, meaning that plastics put into the ocean can float around and cause problems for many years. Animals of all kinds die as a result of being tangled in plastic debris, and many animals mistake plastic for food and die as a result of ingesting it.

Ocean dumping also impacts humans in a variety of detrimental ways. Floating debris can damage recreational and fishing vessels. In 2009, U.S. recreational boaters suffered 165 accidents, 13 deaths, 56 injuries, and damages of $1,469,118.72. Refuse that washes ashore pollutes beaches and deters tourists, thereby reducing tourism revenues. Relict fishing gear and "ghost" crab pots cause economic losses in fishing. The Asia-Pacific Economic Cooperation estimates that the 21 Pacific Rim countries in APEC lost $1.265 billion because of marine debris in 2008.

Legislation is aimed at restricting the ocean dumping that takes place in all navigable waters of the United States. Annex V of the Marpol Treaty designates how far out in the ocean certain materials must be dumped, and makes it illegal to dump any plastic anywhere in U.S. waters. While this type of legislation has undoubtedly had a positive effect, it represents only a small step toward alleviating the problems associated with ocean dumping. There are several other things being tried to improve the situation, the most promising being the reduction of plastics use and recycling. Improvement may also come from the development and use of naturally degradable plastics. (Students can experiment with the degradation of plastics and other materials in Activity 18.)

The ocean can be viewed as an integral component of the global environment. It affects temperatures and climatic conditions worldwide and is largely responsible for making Earth a habitable planet. The ecologically significant characteristics of the ocean stem from the delicate balance of its many natural systems. Human activities that disrupt this balance threaten to alter the environmental conditions upon which we depend for survival. Therefore, knowledge about the ocean and informed decision making regarding its use are essential for the maintenance of a life-sustaining planet.

Constructing a Wave Tank

Constructing a wave tank is not overly difficult, although experience working with Plexiglas or building other projects will make the job easier. Give yourself plenty of time to work slowly and carefully. You might want to begin construction one to two weeks before scheduled classroom use to allow for construction and practice. Read all directions before beginning the assembly. You may also want to have extra materials on hand in case problems arise. (Note: Because the U.S. building industry still uses the English system of measurement, the instructions that follow will be in English units.)

Materials for Constructing a 9-foot Wave Tank

- nine pieces ¼ in. Plexiglas measuring 12 in. × 36 in. (S1A through S3B and B1 through B3)
- two pieces ¼ in. Plexiglas measuring 12 in. × 12 ¼ in. (E1 and E2)
- one piece ¼ in. Plexiglas measuring 11 in. × 15 in. (Paddle/Wave Generator)
- one piece ¼ in. Plexiglas measuring 11 in. × 24 in. (Beach)
- one piece ¼ in. Plexiglas measuring 4 in. × 4 in. (Prop 1)
- one piece ¼ in. Plexiglas measuring 2 in. × 4 in. (Prop 2)
- six transparent acrylic joiners: four @ 12 in. and two @ 11 ½ in.
- four acrylic hinges
- one tube acrylic cement
- power drill with bit size appropriate for drain plug
- one tube silicone sealant
- carpenter's square
- drain plug (about ½ in. diameter)
- overhead transparencies (three)

Procedure

1. The wave tank is assembled in three sections that may be taken apart for storage. (See **Figure 1**.)

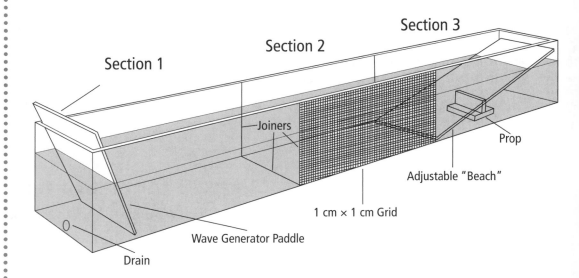

Figure 1
The assembled wave tank

The Plexiglas should be purchased precut to the specified sizes. Once you have assembled the necessary materials, arrange them as shown and label each piece of Plexiglas according to **Figure 2**.

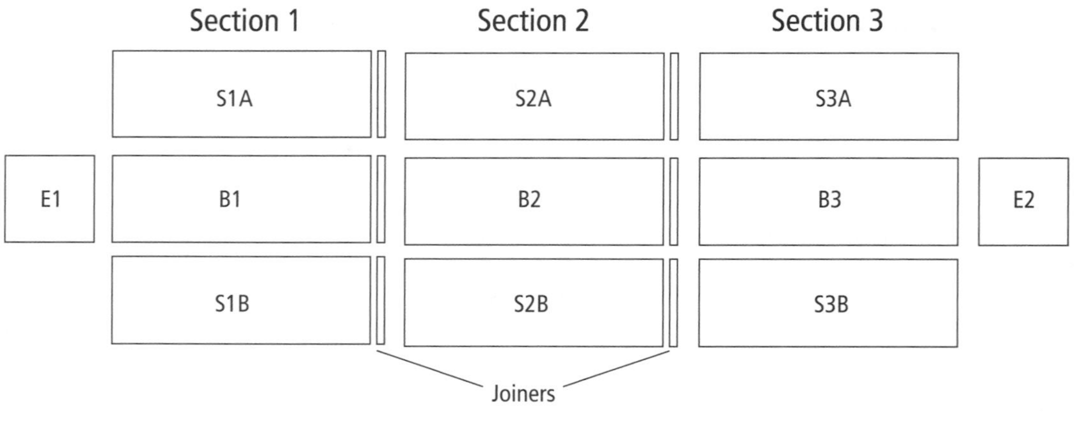

Figure 2

Section 1	Section 2	Section 3
S1A	S2A	S3A
B1	B2	B3
S1B	S2B	S3B

E1 ... E2

Joiners

2. Into piece E1, carefully drill the drain hole. Take care when doing this so as not to crack the Plexiglas. The plug should fit snugly into the hole to prevent leakage. Position the hole as shown in **Figure 3**.

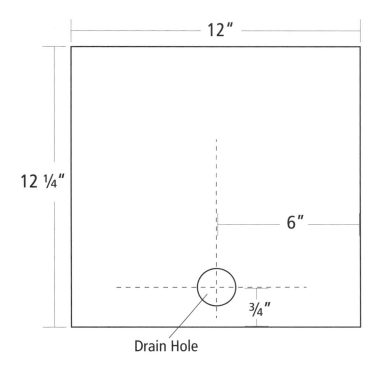

Drain Hole

Figure 3

3. Carefully glue piece S1A to B1, using the carpenter's square to ensure the pieces are square to one another. Put a line of the acrylic cement along the edge of piece S1A and glue the two pieces together as shown in **Figure 4**. Important: Make sure the two pieces are square to one another and the ends of the two pieces are even.

Once S1A and B1 are firmly glued together, glue S2A to B2, and S3A to B3, as described above.

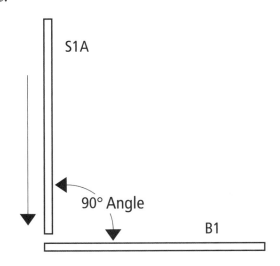

S1A

90° Angle

B1

Figure 4

4. Carefully glue piece S1B to S1A/B1, using the carpenter's square to ensure the pieces are square to one another. Put a bead of the acrylic cement along the edge of piece S1B and glue the two pieces together as shown in **Figure 5**. Important: Make sure the three pieces are square to one another and the ends of the three pieces are even.

Once S1B and S1A/B1 are firmly glued together, glue S2B to S2A/B2, and S3B to S3A/B3, as described above.

Figure 5

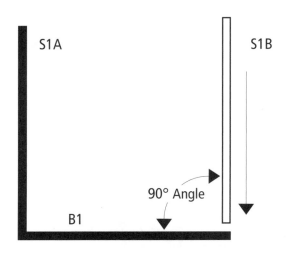

5. Carefully glue piece E1 to S1A/B1/S1B, using the carpenter's square to ensure the pieces are square to one another. Put a bead of the acrylic cement along the edge of pieces S1A/B1/S1B and glue the four pieces together as shown in **Figure 6**. Glue piece E2 to S3A/B3/S3B as described above.

Figure 6

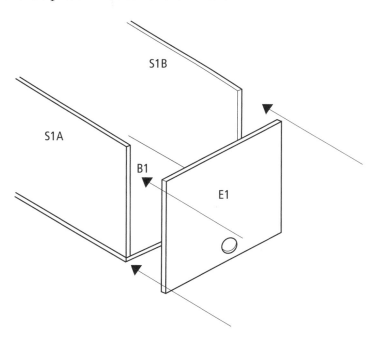

6. Carefully glue a pair of hinges to each of the two pieces of Plexiglas indicated below and label them "Beach" and "Paddle." Put a bead of the acrylic cement along one side of the hinge and glue the hinge to the Plexiglas as shown in **Figure 7**. Important: Be careful not to get glue into the hinging mechanism.

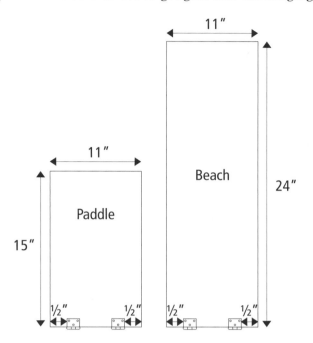

Figure 7

7. Carefully glue the hinges on the "Paddle" to piece B1. Put a bead of the acrylic cement along the side of each of the two hinges and glue the hinges to the Plexiglas as shown in **Figure 8**. Make sure to center the "Paddle" within the tank, leaving approximately ¼" between the "Paddle" and the sides of the tank. Important: Again, be careful not to get glue into the hinging mechanism. Repeat this procedure to attach the "Beach" to B3.

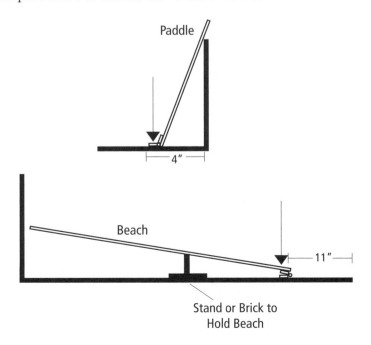

Figure 8

8. Using a bead of acrylic cement, glue the transparent joiners to the open end of Section 1 and to one end of Section 2 as shown in **Figure 9**. (See **Figure 2** to identify the correct placement of the joiners.)

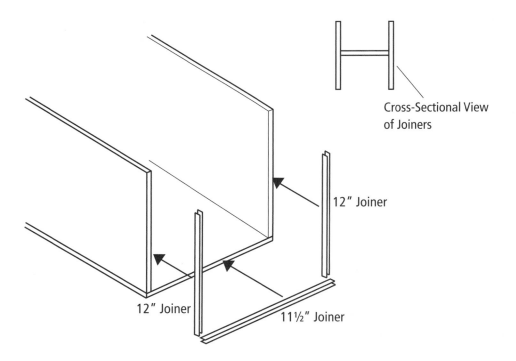

Cross-Sectional View of Joiners

12″ Joiner

12″ Joiner

11½″ Joiner

Figure 9

9. To make the prop for the adjustable beach, glue the "Prop 2" piece to the "Prop 1" piece as shown in **Figure 10**. Use the carpenter's square to ensure the two pieces are square to one another.

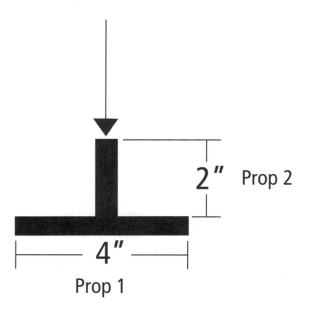

2″ Prop 2

4″

Prop 1

Figure 10

10. After all pieces have had a chance to set (about a day or so), put the wave tank on a counter, preferably with the drain positioned over a sink or close to a sink. This will make it easier to fill and drain the tank.

11. To assemble the three sections of the wave tank, simply fill the channels of the joiners on the end of one of the pieces of the tank with the silicone sealant. (See **Figure 11**.) Follow the directions for the application of the sealant on the package. You do not need to use an excessive amount.

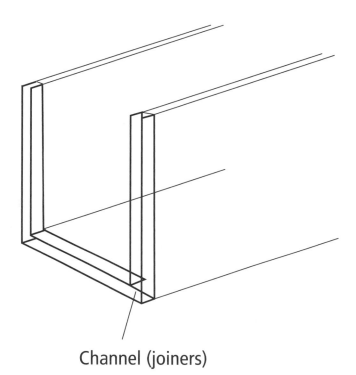

Channel (joiners)

Figure 11

12. Slide the wave tank pieces tightly together, fitting the end of the nonchanneled piece of Section 2 into the channel on the end of Section 1. The sealant may squeeze out of the channel when the tank is properly connected. Repeat this procedure to join Section 3 with Sections 1 and 2. Important: The sealant should be allowed to set overnight before filling the tank with water.

13. Fill the tank with water to a depth of about 15 or 20 cm. Use a red wax marking pencil to note the locations of any leaks. Drain the tank and use the silicone sealant to seal any leaks. Allow the sealant to set overnight before refilling the tank and using it with students.

14. Create three transparent grids by transferring a 1 cm × 1 cm grid from a lab manual to overhead transparencies. Place the grids on the side of the tank, securing them to Section 2 with removable transparent tape as in **Figure 1**.

15. After use, the tank may be disassembled for storage. Drain the tank and pull apart the three sections. Remove any excess sealant from the channel and from the ends of the tank.

Resources

This resource material was compiled by Project Earth Science staff, consultants, and participants, and by the NSTA Press editors. It is not meant to be a complete representation of resources in oceanography, but it will assist teachers in further exploration of this subject.

American Geophysical Union (AGU)

AGU is a professional association of researchers in geosciences. AGU disseminates scientific research on several geosciences topics, including oceanography. Its education department provides multiple resources for teachers.

AGU
Member Service
2000 Florida Avenue N.W.
Washington, DC 20009-1277
tel. 800-966-2481
www.agu.org/education

American Society of Limnology and Oceanography (ASLO)

ASLO is a professional organization for researchers and educators in aquatic science studying both lakes and oceans.

ASLO Business Office
5400 Bosque Boulevard, Suite 680
Waco, Texas 76710-4446
tel. 800-929-2756
www.aslo.org

Bridge: Ocean Education Teacher Resource Center

Bridge provides current marine science information and data, supports researchers doing outreach, and bridges between education and research communities.

Bridge
Virginia Institute of Marine Science
College of William and Mary
P.O. Box 1346
Gloucester Point, VA 23062
tel. 804-684-7000
www.vims.edu/bridge

Centers for Ocean Science Education Excellence (COSEE)

COSEE is a network for connecting educators and scientists to further ocean literacy.

COSEE Central Coordinating Office
Office of Marine Programs
University of Rhode Island
Narragansett, RI 02882
tel. 401-874-6211
e-mail: cosee@gso.uri.edu
www.cosee.net

Environmental Protection Agency (EPA)

The EPA's website links to water-related sites designed for students.

Environmental Protection Agency
Student Center
US EPA Region 5
77 W. Jackson Blvd.
Mail Code P-19J
Chicago, IL 60604
tel. 313-353-6353
www.epa.gov/students/water.html

Estuaries.gov

This is the educational site for the National Oceanic and Atmospheric Administration's (NOAA's) National Estuarine Research Reserve System. The site includes information for students about estuaries, videos, virtual field trips, data from the System-wide Monitoring Program, and other resources about estuaries.

http://estuaries.gov

Great Explorations in Math and Science (GEMS)

Lawrence Hall of Science publications relevant to oceanography for middle school include *Only One Ocean*; *Ocean Currents*; *Convection, A Current Event*; and *Discovering Density*. These curriculum guides describe hands-on activities that complement activities in this volume.

GEMS
University of California, Berkeley
Lawrence Hall of Science #5200
Berkeley, CA 94720-5200
tel. 510-642-7771
e-mail: gems@berkeley.edu
www.lhsgems.org

Long Term Ecological Research (LTER) Network

The LTER network has several sites along the coast at which scientists do research on ecological processes to compare sites across time, space, and environments.

Long Term Ecological Research (LTER) Network Office
UNM Dept of Biology, MSC03 20201
University of New Mexico
Albuquerque, NM 87131-0001
tel. 505-277-2551
e-mail: office_support@lternet.edu
www.lternet.edu

Monterey Bay Aquarium

A west coast aquarium with numerous online educational resources: webcams, podcasts, video clips, lessons, and a seafood watch list.

Monterey Bay Aquarium
886 Cannery Row
Monterey, CA 93940
tel. 831-648-4800
www.montereybayaquarium.org

Monterey Bay Aquarium Research Institute

Marine research projects based out of Monterey Bay, California, include physical oceanography, ocean floor geology, and marine science research. The website has videos of scientists doing research, as well as videos of marine organisms.

Monterey Bay Aquarium Research Institute
7700 Sandholdt Road
Moss Landing, California 95039-9644
www.mbari.org

NASA Earth Observations

Satellite data on oceans, atmosphere, energy, land, and life are displayed as interactive maps.

http://neo.sci.gsfc.nasa.gov/Search.html

National Aquarium in Baltimore

The National Aquarium in Baltimore seeks to inspire people to enjoy, respect, and protect marine environments and their organisms. The website provides information about museum programs, and supplies fact sheets about marine topics at three reading levels.

National Data Buoy Center

This website displays oceanographic and meteorological data from more than 1,000 buoys in the Pacific, Atlantic, and Indian oceans.

National Estuarine Research Reserve System (NERRS)

This program, part of NOAA, is a network of exemplary estuaries protected for long-term research, education, and stewardship.

National Estuary Program (NEP)

The NEP, within the EPA, is a set of significant estuaries with comprehensive plans to address threats by pollution, development, or overuse. The NEP publishes a Coastal Condition Report for all of the 28 estuaries combined, and a State of the Bay report for individual ones.

National Geographic Education

National Geographic's website for teachers and students has online activities, downloadable lesson plans, printable maps, and resource material available for purchase. You can filter resources by topic, grade level, and type.

National Aquarium in Baltimore
501 E. Pratt Street
Baltimore, MD 21202
tel. 410-576-3800
www.aqua.org

National Data Buoy Center
1007 Balch Blvd.
Stennis Space Center, MS 39529
tel. 228-688-2805
e-mail: webmaster.ndbc@noaa.gov
www.ndbc.noaa.gov

NERRS
Estuarine Reserves Division, N/ORM5
Office of Ocean and Coastal Resource Management
NOAA Ocean Service
1305 East West Highway
Silver Spring, MD 20910
tel. 301-713-3155
e-mail: membership@nsta.org
www.nerrs.noaa.gov

U.S. Environmental Protection Agency
Office of Water (4101M)
1200 Pennsylvania Avenue, N.W.
Washington, DC 20460
tel. 202-566-1730
e-mail: ow-owow-internet-comments@epa.gov
http://water.epa.gov/type/oceb/nep

National Geographic Education
1145 17th Street, NW
Washington, DC 20036
http://education.nationalgeographic.com/education/edu

National Ocean Service (NOS)

Part of the National Oceanic and Atmospheric Administration, the NOS translates science, tools, and services into action, working toward healthy coasts and healthy economies. The education team supplies activities, case studies, tutorials, and lesson plans online. Tutorials for students teach about currents, estuaries, tides, pollution, etc.

National Oceanic and Atmospheric Administration (NOAA)

NOAA is the federal agency that protects, preserves, and manages coastal waters and oceans. NOAA includes research, stewardship, and education components. NOAA's Ocean page provides information about coral reefs, tides and currents, buoys, marine sanctuaries, estuaries, diving, oil and chemical spills, and links to marine organizations within and outside of NOAA. Also included are listings of NOAA publications and products.

National Science Teachers Association (NSTA)

A member-driven organization for K–12 science teachers that holds conferences and publishes books and journals. The NSTA Science Store sells many resources on water and ocean studies.

National Sea Grant Office

The National Sea Grant Office is a program of the National Oceanic and Atmospheric Association via universities to serve citizens in coastal communities with research and its applications. The website is a repository of educational publications, information on regional sea grant programs, and funding opportunities.

National Ocean Service
Communications and Education Division
NOAA's National Ocean Service
SSMC4, Room 13317
1305 East-West Hwy
Silver Spring, MD 20910
tel. 301-713-3010
e-mail: oceanserviceseducation@noaa.gov
http://oceanservice.noaa.gov

NOAA
1401 Constitution Avenue, NW
Room 5128
Washington, DC 20230
tel. 202-482-6090
www.noaa.gov/ocean.html

National Science Teachers Association
1840 Wilson Boulevard
Arlington, VA 22201-3000
tel. 703-243-7100
www.nsta.org

National Sea Grant Office
NOAA/Sea Grant, R/ORI
131 5 East-West Highway
SSMC-3, Eleventh Floor
Silver Spring, MD 20910
tel. 301-734-1066
www.seagrant.noaa.gov

Ocean Literacy
This is the website for the network of scientists, educators, and policy makers who developed the "Ocean Literacy Framework" and the "Ocean Literacy Scope and Sequence."

http://oceanliteracy.wp2.coexploration.org

Project Oceanology
Project Oceanology is a nonprofit marine science and environmental education organization in Connecticut. Project Oceanology provides boats, oceanographic equipment, a waterfront laboratory, instructional materials, and staff to provide firsthand, on-the-water experiences.

Project Oceanology
Avery Point
1084 Shennecossett Road
Groton, CT 06340
tel. 860-449-8008
www.oceanology.org

Teachers' Domain
Teachers' Domain is a searchable website of short digital media clips from public broadcasting and partners.

www.teachersdomain.org

The Oceanography Society
This is an organization dedicated to outreach on oceanographic topics.

The Oceanography Society
P.O. Box 1931
Rockville, MD 20849-1931
www.tos.org

Water Resources Division, U.S. Geological Survey
The U.S. Geological Survey provides data about the nation's water resources. The Education portion presents information on many aspects of water, along with pictures, data, maps, and an interactive center.

Water Resources Division
U.S. Geological Survey
12201 Sunrise Valley Drive
Reston, VA 20192
http://water.usgs.gov/education.html

Windows to the Universe
Windows to the Universe is an Earth science website for students, in three reading levels in both English and Spanish.

www.windows2universe.org

Woods Hole Oceanographic Institution
This is a private, independent, not-for-profit corporation for research and higher education in ocean science. The website includes an education section with materials and programs for teachers, and information about marine sciences and marine careers for students.

Woods Hole Oceanographic Institution
Information Office
Co-op Building, MS #16
Woods Hole, MA 02543
tel. 508-289-2252
www.whoi.edu/home

Alfredo L. Aretxabaleta

Alfredo L. Aretxabaleta is a physical oceanographer working in Woods Hole, Massachusetts. He received a BS in marine sciences from the University of Las Palmas de Gran Canaria in Spain in 1998, and a PhD in marine sciences from the University of North Carolina at Chapel Hill (UNC–CH) in 2005. While at UNC-CH, he was part of the implementation of a Coastal Ocean Observing System for the Southeast United States. He then worked as a postdoctoral investigator at the Woods Hole Oceanographic Institution on the physical processes controlling harmful algal blooms ("red tides").

He is currently working at the Institut de Ciències del Mar (Barcelona, Spain) in the development and improvement of the SMOS (Soil Moisture and Ocean Salinity) satellite of the European Space Agency. SMOS is the first satellite measuring salinity from space and introduces a new technological approach to observe the ocean. Aretxabaleta is also working with the United States Geological Survey in Woods Hole on sediment transport problems in the coastal ocean. He uses both ocean circulation models and observations to improve our understanding of the coastal and open ocean. His main interest is the implications of the physical processes in the ocean for problems of interdisciplinary nature such as the behavior of biological organisms or the transport of pollutants and/or sediment.

Aretxabaleta has collaborated in teaching introductory level classes at UNC–CH on physical oceanography and the modeling of marine and Earth systems. He was also involved in the book *Enseñanza Práctica de Conceptos de Oceanografía Física* of the Oceanography Society.

Gregg R. Brooks

Gregg R. Brooks is a professor of marine science with Eckerd College in St. Petersburg, Florida. He received his Bachelor's Degree in geology from Youngstown State University in 1977. He received his Master's and PhD degrees in marine science, specializing in marine geology, from the University of South Florida in 1981 and 1986 respectively.

Brooks's research interests include the study of sediments and sedimentary processes, and the geologic development of open marine and coastal systems. Brooks's current research interests include the study of sediments and sedimentary processes, and the geologic development of open marine and coastal systems. He has worked on a variety of projects ranging from west Florida to south Australia, and he is the author of over 150 publications. In recent years, Brooks has been involved with numerous projects focusing on the Gulf of Mexico and the Caribbean basin. Projects include long-term studies of the development of the west Florida coast and inner shelf, and the sedimentary development of the Tampa Bay and Charlotte Harbor estuaries. Brooks is currently focused on detecting individual events in the sedimentary record, an offshoot of an ongoing study of the sedimentary development and the record of anthropogenic impacts in the U.S. Virgin Islands.

Nancy W. West

Nancy W. West is a geoscience educator in Fort Collins, Colorado. She received an AB in geology from Princeton University in 1979 and an MAT from the University of North Carolina at Chapel Hill (UNC–CH) in 1991. Her studies include graduate coursework in geology, and field and lab analysis of a volcanic tuff from the Mojave desert.

West has taught geology at UNC–CH, chemistry and geology in Durham, North Carolina, high schools, geology for Thomas Nelson Community College, and science methods courses at Duke University and the College of William and Mary (W&M).

In addition to teaching, West has developed Earth science curricula for a middle and high school National Science Foundation project and as the Williamsburg–James City County (Virginia) Public Schools' Science Curriculum Coordinator. Recently, she has been the Curriculum Specialist on the Virginia Demonstration Project at W&M, a project to enhance middle school students' interest in STEM careers, using problem-based learning, which features LEGO robots.

Upon moving back to Colorado in 2009, West started a consulting company, Quarter Dome Consulting, LLC. She is particularly interested in helping teachers set up projects involving field studies, spatial analysis, and the thoughtful use of technology. Her projects involve curriculum and professional development.

Index